Non-Motorized Transport Integration into Urban Transport Planning in Africa

What challenges do pedestrians and cyclists face in cities of the developing world? What opportunities do these cities have to provide for walking and cycling? Based on in-depth research conducted in Cape Town (South Africa), Dar es Salaam (Tanzania) and Nairobi (Kenya), this book explores these questions by presenting work on walking and cycling travel behaviour, the status of road safety in these cities, as well as an analysis of the infrastructure for walking and cycling, and the workings of the institutions responsible for planning for these modes. The book also presents case studies relating to particular opportunities and challenges, such as the development and evaluation of 'walking bus' interventions, and the opportunities micro-simulation of pedestrian interventions offers within a data-scarce environment.

Non-motorized Transport Integration into Urban Transport Planning in Africa demonstrates that transport and urban planning remains situated in a logic of automobile-dependent transport planning and global city development. This logic of practice does not pay adequate attention to walking and cycling. The book argues that a significant shift in both policy as well as political commitment is needed in order to prioritize walking and cycling as strategies for sustainable transport policy in urban Africa.

This book will be a key text for practitioners and policy makers working in planning, transport policy and urban development in Africa, as well as students and scholars of African studies, development studies, urban geography, transport studies and sustainable development.

Winnie V. Mitullah is Associate Research Professor of Development Studies based at the Institute for Development Studies, University of Nairobi, Kenya.

Marianne Vanderschuren is Associate Professor at the University of Cape Town, South Africa.

Meleckidzedeck Khayesi is a teacher by profession, conducting research in Human Geography, with a focus on transport and road safety.

Transport and Society
Series Editor: John D. Nelson

This series focuses on the impact of transport planning policy and implementation on the wider society and on the participation of the users. It discusses issues such as gender and public transport, travel for the elderly and disabled, transport boycotts and the civil rights movement, etc. interdisciplinary in scope, linking transport studies with sociology, social welfare, cultural studies and psychology.

Non-motorized Transport Integration into Urban Transport Planning in Africa

Edited by Winnie V. Mitullah, Marianne Vanderschuren and Meleckidzedeck Khayesi

LONDON AND NEW YORK

First published 2017 by Routledge

2 Park Square, Milton Park, Abingdon, Oxfordshire OX14 4RN
52 Vanderbilt Avenue, New York, NY 10017

Routledge is an imprint of the Taylor & Francis Group, an informa business

First issued in paperback 2019

British Library Cataloguing in Publication Data
A catalogue record for this book is available from the British Library

Library of Congress Cataloging in Publication Data
Names: Mitullah, W. V., editor. | Vanderschuren, Marianne Johanna Wilhelmina Antoinette,
1966- editor. | Khayesi, Meleckidzedeck, editor.
Title: Non-motorized transport integration into urban transport planning in Africa / edited by
Winnie Mitullah, Marianne Vanderschuren and Khayesi Meleckidzedeck.
Description: Abingdon, Oxon ; New York, NY : Routledge, 2017. | Series: Transport and society
Identifiers: LCCN 2016059091| ISBN 9781472411402 (hardback) | ISBN 9781315598451 (ebook)
Subjects: LCSH: Urban transportation—Africa—Planning. | Cycling—Africa. |
Pedestrians—Africa.
Classification: LCC HE311.A4 N66 2017 | DDC 388.4—dc23
LC record available at https://lccn.loc.gov/2016059091

ISBN: 978-1-4724-1140-2 (hbk)
ISBN: 978-0-367-21902-4 (pbk)

Typeset in Times New Roman MT Std
by diacriTech

To all walkers and cyclists in Africa, and all those who are working hard to improve their conditions.

Contents

Figures

Tables

Contributors

Jennifer Baufeldt completed her MSc Civil Engineering degree at the University of Cape Town. During her time as a master's student, she was involved in several projects, including projects for the United Nations Environmental Programme and the National Department of Transportation, all focusing on improving the situation for non-motorized transport. Her main interests are sustainable transport engineering in urban areas, improving equity and social inclusion in developing countries. Jennifer is currently undertaking her PhD at the University of Cape Town, within the field of sustainable transportation under the supervision of Professor Marianne Vanderschuren.

Roger Behrens is the Director of the Centre for Transport Studies and the African Centre of Excellence for Studies in Public and Non-Motorised Transport, and is an Associate Professor in the Department of Civil Engineering at the University of Cape Town. He holds a Bachelor of Arts (1986), a Master of City and Regional Planning (Distinction) (1991) and a PhD (2002) from the University of Cape Town. His current research activities relate to the integration and improvement of paratransit services; the dynamics and pace of changing travel behaviour; the use of transport systems by pedestrians; and the urban form prerequisites for viable public transport networks.

Edward Beukes received a BSc Eng, MSc Eng and PhD from the University of Cape Town. He is a highly qualified engineer with 14 years' experience in transport planning in the public, private and academic sectors, with a project portfolio spanning planning and design, construction, project management, policy development and advanced research and analysis across a range of disciplines in transportation engineering. In his current role at the City of Cape Town Municipality he focuses primarily on transport planning, transport systems analysis and project management. He has contributed widely to transport and land use policy development in Cape Town, in particular looking at urban-tolling, public transport fares and tariffs, travel demand management, public transport, emissions modelling and policy development, parking management and strategic regional land use planning.

Hannibal Bwire is Senior Lecturer, Department of Transportation and Geotechnical Engineering, University of Dar es Salaam. His current research interests include school travel planning, transport systems analysis and planning, travel behaviour analysis, planning methodologies for improvement of transport systems in informal settlement, travel demand management and transport data collection methods. He graduated with a PhD in 2007, which focused on the application of urban transport planning models in developing countries, through a sandwich programme between the University of Dar es Salaam and Darmstadt University of Technology, Germany.

Patrick Chacha is a holder of an MSc in Highway Engineering (Urban Transportation and Planning) from the University of Dar es Salaam, working under the Government of Tanzania as District Engineer at the Kaliua District Council. Patrick is a Professional Engineer researching urban transportation and planning in Dar es Salaam. His publication concentrates on urban transportation and planning (school travel planning), non-motorized transportation (walking school bus and cycle train), children's independent mobility and modelling of mode choice in schoolchildren.

Ezra Goldman holds a bachelor's in Cultural Anthropology from Reed College, Portland, Oregon, and a master's in City Planning from MIT, Cambridge, Massachusetts. He contributed to this article while completing coursework for a PhD at the University of Copenhagen, where he researched the motivating factors for Danes to chose to ride a bicycle for transportation. He has worked with Fortune 100s, governments and non-profit organizations internationally to innovate new products and services based on a deep understanding of human needs. He is currently in Silicon Valley building a new carshare service called Upshift.

Gail Jennings has an MA in Linguistics from the University of Stellenbosch, South Africa, in which she examined the role of metaphor and myth in constructing the narrative of the private vehicle lifestyle. She has worked as a researcher for 25 years and in 2008 founded the transport policy journal Mobility. Jennings has published and presented widely on utility cycling, has planned the walking and cycling network for three Bus Rapid Transit (BRT) systems in South Africa, and undertaken a feasibility study for public bicycle systems in Cape Town. She currently publishes the Cape Town Bicycle Map, and is working toward her PhD, contemplating the shifting ideas and ideologies that affect utility cycling behaviour in South Africa.

Rahul Jobanputra has over 30 years of experience in civil engineering, transport design and construction projects. He holds a BSc (Hons) from the United Kingdom and master's and PhD qualifications in Transport Studies/

Planning from the University of Cape Town. During his career he has worked for private consultancies, contractors and government. He is currently employed by the City of Cape Town as Head of Transport Planning and Policy Development in Transport for Cape Town's Transport Planning Department.

Meleckidzedeck Khayesi is a teacher by profession and conducts research in Human Geography, with a focus on transportation. He holds a PhD in Transport Geography from Kenyatta University. He has twenty-seven years of work experience that straddles global health policy development and programming, research and university teaching. He currently serves as a member of the editorial board of the *Journal of Transport Geography*.

Todd Litman is founder and executive director of the Victoria Transport Policy Institute in Canada. It is an independent research organization dedicated to developing innovative solutions to transport problems. His work helps expand the range of impacts and options considered in transportation decision-making, improve evaluation methods and make specialized technical concepts accessible to a larger audience. His research is used worldwide in transport planning and policy analysis. He has worked on numerous studies that evaluate transportation costs, benefits and innovations. He authored the "Online TDM Encyclopedia," a comprehensive Internet resource for identifying and evaluating mobility management strategies. He has worked as a research and planning consultant for a diverse range of clients, including government agencies, professional organizations, developers and non-governmental organizations. He has worked in more than two dozen countries, on every continent except Antarctica.

George Makajuma has worked with the African Development Bank in Nairobi as a Transport Engineer since 2011, overseeing the Kenyan transport portfolio and regional sector operations in Seychelles and at the East African Community. He has over 15 years of professional experience as a transport engineer working with international civil engineering and development consulting organizations, mainly on roads and transport infrastructure development in a number of African countries. He has provided consulting and technical advisory services in a number of African countries. Earlier in his career, he also lectured university students in transportation planning courses.

Estomihi Masaoe is a Senior Lecturer in the Department of Transportation and Geotechnical Engineering at the University Dar es Salaam. He obtained his undergraduate degree from the University of Dar es Salaam in 1984 in the area of civil engineering. He studied highway engineering (MSc) at the University of Strathclyde, Glasgow and was awarded a PhD degree of the University of Dar es Salaam for his research on safety of un-signalized road junctions in the year 2000. He has taught transportation engineering courses at the University of Dar es Salaam and carried out consultancy and research in transportation engineering especially in the area of road safety.

Winnie V. Mitullah is Associate Research Professor of Development Studies based at the Institute for Development Studies, University of Nairobi. She has a BA and MA in Government from the University of Nairobi, and a PhD from the University of York, United Kingdom. Her background is in political science and public administration specializing in local governance, in particular policies and regulations relating to provision and management of urban services. Her most recent publications in transport are two co-authored chapters in the book *Paratransit in African Cities: Operations, Regulations and Reform.* She is the regional director for Afrobarometers Surveys.

Patrick Muchaka is currently a fulltime doctoral candidate at Stellenbosch University, specializing in pedestrian safety. He was previously the Programme Manager at the Global Road Safety Partnership South Africa from January 2013 to June 2015. He still, however, remains an active member of the GRSP ZA team. He graduated with a master's degree in Transport Studies from the University of Cape Town in 2012. Patrick did his undergraduate studies at the University of Zimbabwe, graduating with a Bachelor of Arts degree in 2000 and an Honours degree in Geography in 2006. He has co-authored a forth-coming literature review journal paper on mobility and access in Sub-Saharan African cities. He has also co-authored a journal paper on child independent mobility in South Africa, published in *Global Studies of Childhood* in 2011.

Japheths Ogendi, PhD, was formerly at Maseno University and currently a Senior Lecturer in the School of Public Health at Mount Kenya University, Kigali, Rwanda. He has conducted research in a wide range of topics in epidemiology of injuries and other non-communicable diseases. He has collaborated with African Center of Excellence in Non-Motorized Transport (ACET) in research on non-motorized transport safety in urban areas. In recent years, Ogendi has emphasized the need to improve safety of pedestrians and pedal cyclists in urban areas with a view to encourage their use as daily modes of transport, to increase physical activity and enhance overall population health. His latest research work is on evaluating public health interventions aimed at improving community health.

Romanus Opiyo is Lecturer in the Department of Urban and Regional Planning, University of Nairobi. He has over ten years of experience in university teaching, research and consultancy with a focus on institutional governance and livelihood. He is a registered lead environmental impact assessor.

Brett Petzer completed undergraduate studies in politics (2007) and architecture (2011) before adopting a bicycle as his primary mode of transport around Cape Town. Ensuing years as an urban cyclist introduced him to issues of inequality, privilege and exposure to risk on South African roads, and the histories behind them. These interests led to a Master's degree in city and regional planning at the University of Cape Town (2016), where his dissertation considered cycling behaviours among three different income groups sharing the same mobility corridor. In 2017, Brett started a PhD in cycling-based mobility services at TU/e, the Eindhoven Technical University.

Marianne Vanderschuren is Associate Professor at the University of Cape Town. She holds a bachelor's degree in Transport Planning and Engineering from Tilburg in the Netherlands, a master's degree in System Engineering, Policy Analysis and Management from Delft, the Netherlands, and a PhD in modelling of Intelligent Transport Systems from Enschede, the Netherlands. She has worked for more than ten years as a researcher at the University of Delft and the Technical Scientific Research Institute in the Netherlands, before joining the University of Cape Town where she, for the past 16 years, translated international best practices in transport planning and engineering into the developing world context. Her main area of expertise is identifying safe, efficient and environmentally friendly transport systems through the use of models and assessment tools.

Eduardo Vasconcellos is an engineer and a sociologist. He holds master's and doctoral degrees in Public Policy from the University of Sao Paulo. He conducted post-doctoral work on urban transport in developing countries at Cornell University, USA. He is currently an advisor for the Brazilian Public Transport Association and for the Bank of Development for Latin America.

Mark Zuidgeest is SANRAL Chair and Associate Professor of transport planning and engineering in the Centre for Transport Studies at the University of Cape Town in South Africa. Trained as a civil engineer, his research and teaching are in the area of sustainable urban transport, an interest that stems from his PhD, completed in 2005 on that topic. Prior to joining UCT in 2013, he had worked for the internationally acclaimed Faculty of Geo-Information Science and Earth Observation (ITC) at the University of Twente (The Netherlands) where he was able, as a transport planning and engineering expert, to link with international scholars in the fields of geosciences, planning and development studies. At the University of Twente, he developed a research portfolio that includes the development of geo-spatial modelling techniques to model, analyze and assess urban transport systems, mostly in developing countries.

Foreword

Transport is critical to the functioning of urban and national systems. Different modes of transport compete and complement each other in the movement of goods and people in urban and national space economies. Non-motorized transport in African cities faces several challenges, with its genesis being the lack of, or inadequate, consideration in urban transport planning due to several historical and contemporary factors. In spite of the increased use of walking and cycling as key modes of transport for improved health and a better environment, non-motorized transport facilities are generally inadequately provided for in several African cities.

Several individuals and institutions contributed and invested in the research and publishing of this book. It presents findings for a better understanding of the travel situation in the three African cities of Cape Town, Dar es Salaam and Nairobi, and accompanying planning for non-motorized transport. The book specifically provides details on walking and cycling travel behaviour, risks facing pedestrians and cyclists, status of infrastructure for walking and cycling, and the workings of institutions with the responsibility of planning for walking and cycling.

The research underlying this book and the process of preparing it advances the value of collaborative research of different scholars and academic institutions in generating knowledge on urban transport in Africa. VOLVO Research and Educational Foundations (VREF) supported collaboration among the Universities of Nairobi, Cape Town and Dar es Salaam, which resulted in a good consolidation and in-depth analysis of key themes on non-motorized transport in three African cities that formed the empirical setting for this book. Non-motorized transport is at the centre of policy on sustainable transport and the contribution of a dynamic group of scholars from African universities to this theme is commendable.

I am confident that this book will be an important resource for researchers and policy makers working on urban transport in Africa and other parts of the world. This book invites us to take a fresh look at urban transport planning in Africa. In particular, it challenges us to systematically include non-motorized transport into urban transport planning in African cities. A key message that runs through the book is the need for a significant shift in both policy planning and political commitment, if we are keen on prioritizing non-motorized transport as part of key strategies for sustainable transport policy in urban Africa. The book is a reference and a reminder that as planners, leaders, citizens and researchers, we have a responsibility to contribute to better planning for non-motorized transport in African cities.

Prof. Peter M. F. Mbithi
Vice Chancellor
University of Nairobi

Preface

What are the opportunities and challenges in promoting walking and cycling around the world? Based on in-depth empirical research conducted in Cape Town, Dar es Salaam and Nairobi, this book answers this question by providing details on walking and cycling travel behaviour, risks facing pedestrians and cyclists, status of infrastructure for walking and cycling, and the workings of institutions with the responsibility of planning for walking and cycling. The book presents examples of transport efforts related to the walking school bus, benefits of simulation of pedestrian measures that can be implemented in African cities, and examples of policy change in favour of walking and cycling. Whereas a huge opportunity exists in sustainable transport policy, the book shows that, except in isolated cases, the planning and governance of transport and urban development is still largely embedded in a logic of automobile dependent transport planning and global city development. This logic of practice does not pay adequate attention to walking and cycling. The book concludes that a significant policy shift is needed if walking and cycling are to be prioritized as strategies for sustainable transport policy in urban Africa.

Winnie V. Mitullah, Marianne Vanderschuren and Meleckidzedeck Khayesi

Acknowledgements

This book is the product of contributions of several individuals and organizations. We thank the Volvo Research and Educational Foundations for providing funding for the research upon which this book is based. We thank the African Centre for Studies in Public and Non-Motorised Transport Director (Roger Behrens) and project leaders (Winnie V. Mitullah and David Mfinanga) for effective leadership. We are grateful to the University of Cape Town, University of Dar es Salaam and University of Nairobi administrations for providing supportive research environments. We greatly acknowledge the hard work of all the researchers and writers who have contributed chapters to this book (see the table of contents for their names). Gail Jennings' attention to the thorough editing of this book is appreciated. We are grateful to Cheryl Wright for assisting with the preparation of this book.

We acknowledge insightful feedback from John Whitelegg, Margaret Grieco, Tim Schwanen, Frederick Nafukho, Adjo Amekudzi, Kwesi Darkoh, Jan Ketil Rød and Jacqueline Klopp.

The staff of Routledge was very supportive in providing guidance and making necessary adjustments to the production timeline as we progressed. We also acknowledge supportive guidance from the staff of what was formerly Ashgate Publishing Limited.

1   Introduction

Challenges and opportunities for non-motorized transport in urban Africa

Marianne Vanderschuren, Gail Jennings, Meleckidzedeck Khayesi and Winnie V. Mitullah

Although walking is the dominant mode in African cities, neither walking nor cycling has been given the attention they deserve, in terms of policy development and practical implementation. African cities remain unsustainably focused on expanding the road networks and increasing motorization, perhaps because of the legacy of colonialism and our contemporary focus on automobile-dependent planning, which see non-motorized transport (NMT) as modes for the poor and motorization as aspirant modes. Where NMT infrastructure has been provided, this is a recent development, to a large extent promoted and supported by international development partners, as part of the global trend toward mitigating climate change and promoting sustainable living. While African cities have seen some improvements for pedestrians, cyclists are now worse off, and mode shares have dropped. Cycling, in particular, attracts minimal attention from both policy makers and urban residents, who view the mode as too risky.

Because of the spatial legacy of apartheid (in South Africa) and colonialism in other African countries, the urban poor continue to live on the outskirts of cities, far from amenities and opportunities. While the poor rely on NMT and public transport to access these opportunities, the distances they travel are not always appropriate for walking and cycling, and lead to extended exposure to personal and traffic safety risk. Thus, provision for non-motorized travel (as a main, as well as a feeder mode) is not only increasingly urgent, but it is also now more challenging given the need to retrofit, and requires large investments in redesigning transport and land-use policies and practices.

The consequences of inadequate NMT planning are a high rate of pedestrian fatalities, poor-quality NMT environments, and an increasing dependence on private car use. As soon as it becomes even slightly financially possible, households in Africa purchase a private vehicle, to reduce the mobility burden they experience. In South Africa, the number of households that own at least one private vehicle increased by 10% between 2003 and 2013 (NHTS, 2003; NHTS, 2013). This has led to debilitating traffic congestion, and rapidly deteriorating urban and air quality.

In the last decade or so, many African states have acknowledged the value of walking and cycling. A recent report by UN Environment

(UN Environment, 2016) shows that most African countries have, in fact, made a start in policy development, although few have policies that focus on NMT alone. Furthermore, the operationalization and implementation of policies, to date, has not yet led to substantive changes in NMT performance. Road fatalities, discomfort and risk remain unacceptably high.

Walking, cycling and public transport are currently part of policy options being pursued toward sustainable transport systems, in view of the challenges of climate change and the Sustainable Development Goals. These modes are also part of the policy discourse around physical and active lifestyles, which aims to find solutions to sedentary lifestyles and associated obesity and cardiovascular diseases. Such policies provide direction while regulations and guidelines focus on implementation and management.

In Cape Town, Nairobi and Dar es Salaam, the cities that form the focus of this publication, broader transport policies and strategies do include some attention to NMT; both Cape Town and Nairobi have, in addition, developed stand-alone NMT policies. Among the vision statements of the case cities are the following intentions:

- To create a safe, cohesive and comfortable network of footpaths and cycling lanes/tracks that include shade
- To develop laws and regulations to ensure prioritization of NMT facilities
- To promote investment in walking and cycling infrastructure
- To connect public transport with walking and cycling facilities
- To influence land-use planning and resettlement patterns to achieve easy access to amenities
- To promote a changed culture that accepts the use of cycling and walking as a means to move around in the city

Nevertheless, despite these policy beginnings, the focus on motorization remains, to a large extent, the transport-planning priority. At the same time, it is of course not too late for African cities to chart an alternative future and pay increased attention to walking, cycling and public transport in both urban and rural areas. Where challenges in NMT infrastructure provision, travel behaviour and policy exist, there are opportunities to improve, particularly as levels of urbanization and motorization are still relatively low. Leveraging these opportunities requires bold policy decisions, political will, and drawing on evidence-based research and sustainable transport measures.

The purpose of this publication

This book focuses on Cape Town, Nairobi and Dar es Salaam, and examines walking and cycling travel behaviour, the risks facing pedestrians and cyclists, the status of infrastructure for walking and cycling, practices in road design and the workings of institutions responsible for planning for walking and cycling. In addition, the book shows that change is possible, by presenting evidence for the possibilities of walking school buses/cycling train, the benefits of micro-simulation of pedestrian safety measures, the benefit of using contextual data for planning purposes, the possibilities provided by infrastructure audits and examples of policy efforts to address the neglect of walking and cycling.

This research accordingly asks questions of a number of the city vision statements presented above, describes the challenges that limit the ability of cities to achieve these visions, and reveals additional gaps that receive insufficient policy attention. These include a concern with the continued focus of NMT as a feeder mode to public transport, the inadequate attention to behaviour change approaches, poor quality or non-existent transport-related data and the uneven quality of infrastructure built. Some of these concerns are highlighted in this chapter, along with key recommendations to translate challenges into opportunities. Each individual chapter deals with these challenges and opportunities in greater depth. This book provides research findings that can be used to support political decision-making and practice in NMT policy and infrastructure development. The book also provides various case studies that offer an indication how to improve planning and assessment.

Method

The book is based on studies undertaken between 2007 and 2015 by the African Centre of Excellence for Studies in Public and Non-Motorized Transport (ACET), funded by the Volvo Research and Educational Foundations (VREF). ACET brings together the Universities of Cape Town, Dar es Salaam and Nairobi. The studies cover the following themes: paratransit operations and regulations, transport systems and travel behaviour, transport planning, practices and governance systems, road safety, and non-motorized transport. This book provides an overview of the non-motorized transport related studies carried out by ACET.

Key findings and recommendations

The key findings in this book are summarized below. Further information on specific elements can be found in the chapters themselves.

NMT is the dominant mode of transport in African cities

In Chapter 2, *Non-motorized travel behaviour in Cape Town, Dar es Salaam and Nairobi* (Marianne Vanderschuren and Gail Jennings), the authors make a clear case that NMT must be taken seriously as a mode. The chapter presents survey findings that show that walking accounts for 99% of all NMT trips in all three cities, while cycling makes up the other 1%. Compared to all modes of transport, walking makes up between 50% and 90% of daily trips in many of the African cities reflected upon in this publication, while cycling makes up a low percentage of trips, varying between 0.05% and 2%. In Nairobi, Dar es Salaam and Cape Town, walking constitutes 73.7%, 70.3% and 46.7%, respectively, of the overall modal split.

The chapter demonstrates not only the importance of NMT in the African context, but also that, despite many decades of concern regarding the poor provision of infrastructure for this major mode, political attention, revenue allocation and facilities provision has not sufficiently shifted towards NMT users.

As other chapters show, NMT users thus continue to bear the inequitable burden of road-crash-related injuries and fatalities and, despite the increasing road congestion and declining urban quality in these cities, and cost of motorized transport, there appears to be insufficient incentive for people to move from motorized to non-motorized modes.

NMT behaviour comprises complex attitudinal dynamics that are linked to age, gender and background, and role players need to understand these in order to provide adequate infrastructure and safety programmes

The nuances of pedestrian behaviour and attitudes, as well as pedestrian-driver interaction, along arterials and freeways, are examined in Chapter 3, *Pedestrian crossing behaviour in Cape Town and Nairobi: observations and implications* (Roger Behrens and George Makajuma). Here the authors report on a series of small indicative studies, undertaken between 2004 and 2010, that observed pedestrian crossing behaviour and collected attitudinal information on selected arterials and freeways in Cape Town and Nairobi, using a variety of methods. It shows that the distribution of points of pedestrian arterial and freeway crossing, in relation to provided crossing facilities, does not follow a sigmoidal curve in Cape Town. Significant numbers of pedestrians were observed to cross arterials and freeways unassisted at small distances from crossing facilities. The relatively greater freeway crossing facility use is likely to be associated with, amongst other factors, a greater perceived risk associated with the greater speed differential on freeways.

In Nairobi, pedestrian crossing (refuge) islands and unsignalized zebra crossings on Jogoo Road were found to be ineffective as a means of simultaneously facilitating safe pedestrian crossing and achieving a significant 'no-discomfort' traffic calming effect on speeding motorists. Significant numbers of pedestrians were observed to cross arterials and freeways unassisted at small distances from crossing facilities, and instances of greater compliance are best explained by the location of the crossing facilities in relation to dominant pedestrian desire lines.

The authors conclude that understanding or estimating pedestrian desire lines and walking trip assignment is more important than understanding detour refusal distances in locating crossing facilities and in attempting to minimize unassisted or illegal crossing patterns. They recommend that walking be routinely included in travel behaviour analysis, when in the past it has been omitted, and treated as a travel mode like any other.

Further, they recommend that infrastructure, such as raised zebra crossings, be considered as a way to improve the safety of pedestrians, while at the same time achieving a level of operational efficiency in the use of limited road space.

Finally, they recommend that education and awareness programmes pay attention to the appreciation of traffic risk, and that for education and awareness campaigns to be most effective, research into variations in attitudes and behaviour across population segments is necessary to identify the most appropriate communication medium and targeting of campaigns.

NMT users are disproportionately involved in road traffic crashes

Chapter 4, *Road safety and non-motorized transport in African cities* (Marianne Vanderschuren and Mark Zuidgeest) shows that African countries have an overall road fatality risk that is substantially higher than the global road fatality risk. This chapter provides an overview of the fatality rates in Africa and African cities, as well as detailed information about fatality levels in Nairobi, Dar es Salaam and Cape Town. It should go without saying that there is a need to reduce the road safety burden in Africa, especially for NMT users. The authors identify vehicle maintenance and speed, as well as substance abuse, as having a significant influence on the road safety risk. Pedestrian crossing behaviour, often caused by security risks and poor land-use planning, is also a major issue.

Reducing the road safety risk on the African continent requires holistic mitigation strategies. NMT specific strategies, typically, focus on the provision of information and education with the aim to change road user attitudes and improve enforcement, with special focus on the reduction of identified illegal behaviour (speeding, alcohol abuse and pedestrian crossings), as well as reduction of the encroachment onto NMT facilities by other modes. To develop these multiple mitigation strategies, disaggregated data needs to be collected and analyzed to create the scientific base for appropriate action.

In Chapter 5, *Types of injuries and treatment of pedestrians admitted to a referral hospital in Nairobi City, Kenya* (Japheths Ogendi), the author notes that a significant proportion of those occupying hospital beds in Kenya are pedestrians, which exerts overwhelming pressure on health-care facility resources.

The chapter examines the age and gender, the types of injuries, and the length of hospital stay treatment of injured pedestrians admitted to a referral hospital in Nairobi. The results show that men between the ages of 15 and 44 years were the most affected. Most of the injuries occurred to the limbs, which has a profound impact on the sustainability of people's livelihoods and their ability to carry out key daily tasks.

In Chapter 6, *Safety of vulnerable road users on a road in Kinondoni municipality, Dar es Salaam, Tanzania*, the author, Estomihi Masaoe, used police records of road traffic crashes occurring in 2008 to characterize fatal road crashes that involved vulnerable road users. In Dar es Salaam, as a whole, vulnerable road users (VRUs) constitute 79% of road fatalities; pedestrian fatalities made up 67% of the total road-based fatalities in 2008. This chapter describes the characteristics of the road crashes involving VRUs, and the interaction between the infrastructure and the users.

There has been progress in the development of NMT policy, guidelines and infrastructure in the case cities

In Chapter 7, *Non-motorized transport infrastructure provision on selected roads in Nairobi* (Winnie V. Mitullah and Romanus Opiyo), the authors establish the availability of NMT infrastructure and facilities in 19 major road corridors in the City County of Nairobi (CCN), and the adequateness of the infrastructure.

Their work shows that the provision of NMT infrastructure has improved, but that the types of infrastructure provided are not uniform and do not fully conform to NMT design principles. Some routes tend to have spacious, good and well maintained NMT infrastructure, while others are either not well maintained or obstructed by other activities. Those who have benefited are pedestrians, while cyclists and people with disabilities are either inadequately provided for or ignored entirely.

The authors note that, largely, the infrastructure provided ignores the way in which NMT is used as a complete mode connecting origins and destinations, and that the discontinuous provision of the facilities point to a conception of NMT as a mere conduit of other modes of transport. As a result, the authors recommend that all new infrastructure and programmes take careful note of the new Nairobi NMT policy (launched in 2015), which aims to facilitate the standard provision of NMT infrastructure in the city. Where standard infrastructure guidelines do not exist, the authors recommend the development or adaptation of such guidelines, and the extension of facilities to all NMT users.

The quality of NMT facilities has a significant impact on NMT trip efficiency and safety; not all NMT infrastructure is effective

Chapter 8, *An investigation into the effects of NMT facility implementations and upgrades in Cape Town* (Jennifer Baufeldt and Marianne Vanderschuren), summarizes the evidence that NMT is beneficial and sustainable for both developing and developed countries, but reiterates that NMT users face various challenges, which reduce the attractiveness of selecting NMT as a mode. The main concern is the inadequate provision of NMT facilities, making NMT trips inefficient and dangerous. The chapter then provides an overview of research conducted at the University of Cape Town, which used case study areas to determine the effects of NMT facilities in Cape Town.

Where NMT implementations have occurred, the study assessed whether NMT facilities hold the same potential benefits for South African NMT users as witnessed in developed countries, despite the possible hindering local context.

Like in Chapter 7, the authors show that more can be done in terms of improving NMT facility implementations in South Africa, in order for local challenges and the needs of NMT users to be better accommodated in a more consistent manner. Recent provisions for pedestrians are implemented more appropriately than recent facilities for cyclists. They recommend that the adoption of South Africa's NMT Facility Guidelines (Vanderschuren et al., 2014) is likely to lead to more successful infrastructure interventions. The focus going forward should be to increase the consistency and quality of the NMT facilities that are maintained and improved upon, as well as increasing the reach and connectivity of the NMT network for both pedestrians and cyclists accordingly.

Local context matters with regard to road categorization and transport infrastructure

In Chapter 9, *Access and mobility: multi-modal approaches to transport infrastructure planning* (Edward Beukes, Marianne Vanderschuren and Mark Zuidgeest), the concern with the high rate of pedestrian deaths is continued, with a focus, this time, on South Africa. South Africa's National Road Safety Strategy Report proposes a number of strategies to address this concern, chiefly:

1 A general improvement of law enforcement measures
2 Enhanced road user education campaigns
3 The expanded implementation of traffic calming schemes

This chapter adds a further strategy, which evolved from a concern with the extent to which road categorization influences the provision of transport infrastructure and, therefore, the danger when rigid road classification is applied, overlooking the needs of NMT. Attempts to address the problem have, generally, tried to recast the categorization problem in more rigid terms, describing roads as being either suited to mobility needs only, or access needs only, instead of the more traditional view that many roads actually support a mix of functions.

The authors recommend a Context Sensitive Design (CSD) as an alternative (interchangeably referred to as Context Sensitive Solutions (CSS)), which determines that certain modes are better suited to a particular set of contextual circumstances than others. Therefore, under a given mix of contextual circumstances, certain modes should be given a higher priority than the rest. Priority then, and by extension context, can be used to determine infrastructural needs.

***Walking buses and cycle trains have the potential to reduce
road congestion, improve independent child mobility and safety,
and encourage physical activity***

Chapter 10, *Implementation and evaluation of walking buses and cycle trains
in Cape Town and Dar es Salaam* (Hannibal Bwire, Patrick Muchaka, Roger
Behrens and Patrick Chacha) starts with the premise that school travel planning is an important, but largely neglected, aspect of the local transport
planning process in Sub-Saharan African cities, and that travel by children
is poorly understood. Because traffic congestion and highway construction
have dominated travel surveys and demand forecasting methods, data collection was limited to, and the travel demand models developed were calibrated
for, motorized trips occurring within the weekday morning peak period when
congestion was generally worst. In many instances only trips to work have
been included.

This chapter describes two school travel initiatives – walking buses and
cycle trains – implemented in Cape Town and Dar es Salaam. Evaluation of
each project revealed that the majority of learners enjoyed walking or cycling
to school, and the majority of parents supported the initiative and would have
continued making use of it, had the project continued.

A key recommendation for those intending to implement walking buses
and cycle trains is, therefore, that the institutional arrangements surrounding
walking buses, and the degree of proactive support provided by the school and
the local municipality, is just as important as the technical questions around
setting up the walking buses and optimizing routes and schedules. Secondly,
local municipalities, and individual schools, need to take responsibility for
such interventions rather than relegating them to volunteers and parents.

***The predominance of unreliable crash data often leads to incorrect
road-safety conclusions and interventions; micro-simulation is able to
go some way toward overcoming this challenge***

Chapter 11, *The use of microscopic simulation modelling techniques to assess
and predict road safety through an analysis of road user and infrastructure interaction in Cape Town* (Rahul Jobanputra) adds to the multi-faceted research
in this publication regarding ways in which to improve pedestrian safety. The
underlying reasons for pedestrian fatality rates are complex, notes the author,
but in resource-poor countries, local authorities and practitioners rely on a
reactive assessment of historic crash data to determine hazardous locations.
However, this data is unreliable, due to recording issues, and it does not contain sufficient detail for comprehensive investigation; the result is a distinct
possibility of inappropriate conclusions.

The author shows successful fatality reductions in countries that use
complementary, alternative evaluation methods and predictive modelling to
proactively assess or predict safety.

Through case studies in Cape Town, his investigation reports on how one such alternative technique, micro-simulation modelling, can be used to provide a better understanding of the interaction between the road user and the infrastructure. Through the innovative use of modelling output, it can be successfully used to evaluate the potential benefits of engineering countermeasures and to provide a comparative safety evaluation of urban infrastructure with different operational characteristics.

Policy intervention and change is not a one-off, centralized event but an iterative process that requires synergy between multiple stakeholders

In Chapter 12, *Institutional framework for walking and cycling provision in Cape Town, Dar es Salaam and Nairobi* (Winnie V. Mitullah and Romanus Opiyo), the authors consider the multiple actors and relationships involved in urban transport policy development and implementation.

Policy is central in promoting walking and cycling. In Africa, only Uganda and South Africa have national NMT policies, although a number of cities, including Nairobi, have city NMT policies, and NMT is mentioned in the majority of national transport policies. Nevertheless, there is the risk that NMT infrastructure promotions, often planned and implemented with the support of development partners, are not grounded in local policy. This makes the outcome of these interventions uncertain.

The authors, therefore, strongly recommend a move away from centralized approaches to transport planning, which too often results in the exclusion of most stakeholders. Instead, they recommend collaboration and coordination to facilitate the mobilization of both resources and actors toward the institutionalization of NMT.

A narrow, 'infrastructure-first' approach has limitations in terms of promoting utility cycling and increasing mode share

Chapter 13, *When bicycle lanes are not enough: growing mode share in Cape Town, South Africa: an analysis of policy and practice* (Gail Jennings, Brett Petzer and Ezra Goldman) looks at the way in which policies and programmes to increase the rate of bicycle transport (or utility cycling) in South Africa place great emphasis on the need to create physical bicycling infrastructure. Although the policies have referenced the need for what the authors refer to as 'soft' infrastructure – the promotion of a 'culture and respect of NMT', supportive law enforcement, vehicle-free days, and incentives, such as showers and bicycle travel allowances – the actual outcome of such policies so far has been the creation of underused, relatively disconnected and inconsistent hard infrastructure.

This chapter first provides an overview of South Africa's national and local 'infrastructure-first' approach, and discusses the way in which this narrow approach, evident in Cape Town until recently, may go some way to explaining the lack of significant mode shift in the city, since the first walking and cycling policy in 2008. The authors contrast engineering-led approaches to understanding cycling, which are prevalent in low-cycling contexts, with approaches drawing on the social sciences, where aspects such as culture, attitudes, behaviours, discourses and beliefs are given greater weight in cycling planning.

Efforts to promote cycling in Cape Town and South Africa have attained insufficient returns, to date, and run a high risk of continued failure without a commitment to a change in direction toward a behavioural approach.

Chapter 14, *Grounding urban walking and cycling research in a political economy framework* (Meleckidzedeck Khayesi, Todd Litman, Eduardo Vasconcellos and Winnie V. Mitullah), the authors argue the need to ground urban walking and cycling research in a political economy framework, which highlights the need for synergy among public transport actors, in particular policy makers, practitioners, researchers and advocates of public transport, to build understanding to support walking and cycling.

The authors recommend that policies take an integrated approach, which gives equal consideration to walking, cycling and mass rapid transportation, rather than a concentration on motorized modes of public transport.

References

National Household Travel Survey (NHTS) 2003. Database.
National Household Travel Survey (NHTS) 2013. Database.
UN Environment. 2016. *Global outlook on walking and cycling 2016.* Nairobi: UN Environment. Available at: http://wedocs.unep.org/handle/20.500.11822/17030
Vanderschuren M., Phayane, S., Taute, A., Ribbens, H., Dingle, N., Pillay, K., Zuidgeest, M., Enicker, S., Baufeldt, J. and Jennings, G. 2014. *NMT facility guidelines, 2014 – policy and legislation, planning, design and operations.* Pretoria: Department of Transport, Pretoria.

2 Non-motorized travel behaviour in Cape Town, Dar es Salaam and Nairobi

Marianne Vanderschuren and Gail Jennings

Introduction

In almost every urban centre in Africa, non-motorized transport (NMT) and walking, in particular, is the dominant mode of travel, driven by financial necessity rather than by sustainability or health motives. The Sub-Saharan African Transport Policy Programme (SSATP), a partnership between 40 African countries, reports that walking currently makes up between 50% and 90% of daily trips (SSATP, 2015); cycling makes up a low percentage of trips, varying between 0.05% and 2%. This has been the pattern in urban Cape Town, Nairobi and Dar es Salaam, at least since NMT travel behaviour data has been collected (Pendakur, 2005; Rwebangira, 2001; Behrens, 2009).

This chapter presents an analysis of recent NMT travel behaviour data in Cape Town, Dar es Salaam and Nairobi, collected in 2010. A comparison between historic data in the three case cities, and the 2010 collected data, reveals that NMT modes are underestimated, due to a traditional focus on commuter travel. In all cities, NMT is the dominant mode, which makes a case for why improvements in the NMT networks and infrastructure provision are essential.

Although the population of the three case cities is similar, the land area is not, which leads to large differences in average densities (see Table 2.1). This has consequences for travel behaviour. Nairobi, for example has, on average, a higher density than Cape Town and Dar es Salaam. The Gross Domestic Product (GDP) is significantly higher in Cape Town. Vehicle ownership per 1,000 population is higher in Cape Town, in line with GDP statistics. Overall, private vehicle ownership in the case cities is low (when compared to, for example, the US, which has a vehicle ownership of around 800/1,000 population, and Europe, with vehicle ownership rates of around 500/1,000 population (Table 2.1).

Key findings of the travel behaviour analysis presented here relate to travel mode, travel time, trip rates (i.e. how many trips people make – also known as levels of mobility or mobility rates) and the sequence or pattern of trips. NMT in this chapter refers specifically to walking and cycling; neither hand-cart use, trolley use nor animal-drawn transport is considered. Further, the

Table 2.1 Urbanization indicators for the three case cities

Performance indicator	Cape Town (CCT, 2012)	Dar es Salaam (URoT, 2012)	Nairobi (KNBS, 2010)
Population (million)	3.740 (2011)[3]	4.365 (2012)	3.138 (2009)
Area (km²)	2,461 (2011)	1,400 (2012)	695.1 (2009)
Average density (pop/km²)	1,520 (2011)	3,133 (2012)	4,515 (2009)
GDP/capita (USD)[1]	6,086.45 (2014)[1]	600.66 (2014)[1]	648.84 (2014)[1]
Vehicle ownership/1,000 pop[2]	165 (2010)[2]	7 (2007)[2]	24 (2010)[2]

Source:
[1]Country values based on www.tradingeconomics.com/country-list/
[2]Country values based on https://en.wikipedia.org/wiki/List_of_countries_by_vehicles_per_capita
[3]Based on South African Census 2011

chapter does not explore the importance of increased walking and cycling in cities; nor does it examine the reasons for the high NMT mode split, or examine the barriers, dangers and challenges to NMT use or infrastructure provision. These matters are dealt with in Chapters 4, 5, 6, 7 and 13.

Method

In 2008, with the establishment of the African Centre of Excellence for Studies in Public and Non-Motorised Transport (ACET), the three partner universities – the University of Cape Town, the University of Dar es Salaam and the University of Nairobi – invested in the collection of baseline data regarding private vehicle, public transport and NMT-related travel behaviour in the three cities.

A questionnaire, informed by literature and local practices regarding travel surveys, was drafted, and a dummy database was developed on the basis of the questions and pre-tested, further tested and refined using the pilot study data (Masaoe et al., 2011). The data itself was collected by an independent market research firm.

Data captured for each household, by means of face-to-face interviews, included age, car ownership, gender, occupation, relationship to the youngest household member, physical characteristics, transport expenditure, household income and type of dwelling. The travel diary for each member who travelled, the trip chain information captured for the previous day, together with their corresponding places visited, modes used, and time lapse for a single complete trip and accompanying members (Masaoe et al., 2011), gives a complete view of the mobility pattern on a single day of travel. The decision regarding the form of travel diary to adopt was made through an experiment conducted in 2008 in Cape Town and Dar es Salaam; the experiment involved three forms of travel diary (trip based, activity based and place based). Results indicated that the place-based instrument was suitable for the cities as it gave the greatest trip and trip stage recall and least recording item non-response error (Behrens and Masaoe, 2009).

Due to the financial resources required for the collection of primary data through travel diaries, it was deemed prohibitive to identify a representative sample size for the number of diaries. In each case city, it was decided to contact a minimum of 2,000 households and collect trip data for each member of these households. Table 2.2 provides an overview of the number of households and diaries collected in each city.

In Cape Town and Nairobi, 2,002 households were surveyed, and in Dar es Salaam 2,009 households. Dar es Salaam has the highest number of diaries per household and the highest number of diaries in total. In Cape Town, 13,608 diaries were kept, the consequence of lower numbers of people per household than in Dar es Salaam. Nairobi has a significantly lower diary rate per household (largely due to lower mobility rates).

The primary data, collected by the ACET team, was supplemented with secondary data. When presenting secondary data, this chapter includes South African data (particularly Cape Town) in greater detail. This reflects the greater availability of secondary information in South Africa, where the policy and regulatory regime requires the collection and analysis of NMT data [for example, South Africa's National Land Transport Act (NLTA) of 2009 (South African Government Gazette, 2009) requires that provinces prepare a provincial land transport framework (PLTF) and that local authorities prepare Integrated Transport Plans (ITP); each requires a full chapter on NMT planning]. Kenya does not have a national framework that requires the gathering of NMT data, although in 2015 the City County of Nairobi developed an NMT policy that will require such data collection, including 'regular pedestrian and cyclists traffic surveys, at specified locations and times, including bicycle parking counts at transit stations, to aid in planning pedestrian and cycling projects and evaluating their benefits'.

Table 2.2 Number of households and diaries in case cities

Cities	Cape Town	Dar es Salaam	Nairobi
Number of households interviewed	2,002	2,009	2,002
Number of trip diaries	13,608	16,231	9,598
Diaries/household	6.80	8.08	4.79
Recorded trips/trip segments	12,839	28,454	11,983

Although not every person in the household makes trips, the data collected for Cape Town, Dar es Salaam and Nairobi include 12,839, 28,454 and 11,983 trip segments, respectively.

Travel behaviour in the case cities before 2010

Over the years, various authors have reported on travel behaviour in the case cities. In this section, their findings are summarized. Figure 2.1 provides an overview of the historic data found for the three case cities: Dar es Salaam, Nairobi and Cape Town.

Dar es Salaam

In Tanzania, the dominance of walking and public transport, the comparative insignificance of trips by private car, and the demise of cycling have been apparent in cities since major sector studies started in the early 1970s (Howe and Bryceson, 2000). Dar es Salaam's mobility rates are low compared to other developing countries outside Sub-Saharan Africa, according to Pendakur (2005). In 2000, the Tanzanian National Team for Non-Motorized Transport reported that more than half of all trips were made on foot (Tembele, 2000). Public transport was next in importance, with cycling as the potentially third most important mode (Howe and Bryceson, 2000; also see Figure 2.1).

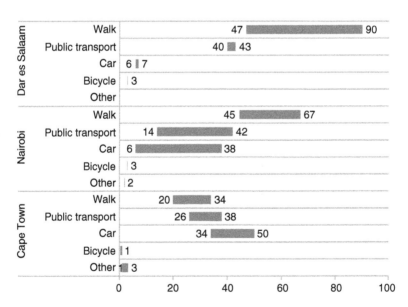

Figure 2.1 Historic data – modal split for all trips (%).

Sources: *Dar es Salaam*: Pendakur, 1994 (cited in Pendakur, 2005); Howe and Bryceson, 2000; Tembele, 2000. *Nairobi*: Howe and Bryceson, 2000; Pendakur, 2005. *Cape Town*: NHTS, 2003; STATSSA, 2011; CCT, 2013; NHTS, 2013 (years refer to data collection period).

Walking accounted for between 47% and 90%, depending on the source used. Cycling, on the other hand, accounts for 3% of the modal split, according to all available sources. People who use public transport were found to take an average of 30 minutes to walk from their residence to the bus stop (Howe and Bryceson, 2000).

The percentage of walking in Dar es Salaam varies significantly, between 47% and 90%. All the other modes display less variance between the different datasets available for the city. In the data referenced above, Howe does not elaborate on modes other than walking. Data collection differences may account for data variance; the other two sources only vary between 47% and 50% walking trips.

Nairobi

In Nairobi, modal share also differs between different historic sources. Pendakur cites household surveys conducted in 1994, which indicated that 47% of all trips were by walking, 1% by bicycles, 42% by public transport, and 10% by private or company car (Pendakur, 2005). Similar travel data was reported at the turn of the 21st century (Rwebangira, 2001; Servaas, 2000). The 1998 Nairobi Long-Term Transport Study (NLTSS), cited by Howe and Bryceson (2000), showed the proportion of cars and vans on the main roads, out of total motorized vehicles, averaged 73%, ranging between 57–88% at different points, and public transport was 19% (range 8–32%). Light and heavy goods vehicles constituted the remaining 8% ranging between 2–15% (Howe and Bryceson, 2000).

Bicycle use has continued to be rare, despite a fairly long history of use: in the early 20th century, government and corporate employers enabled employees to purchase bicycles, and provided facilities between places of work and home (Khayesi et al., 2010; Howe and Dennis, 1993; Howe and Bryceson, 2000). Use and stakeholder surveys are unanimous in identifying a fear of road crashes as the main reason for the decline in use (Howe and Dennis, 1993; Howe and Bryceson, 2000).

The different data sources available for Nairobi show varying modal splits (see Figure 2.1). Walking accounted for 44.6% in 1973, increased to 67% in 1990 and has stabilized at around 47% in the period between 1994 and 2002. The use of public transport has seen a steady growth in Nairobi, from 14% in 1973 to 42% in 2002. Private car use was 38% in 1973, and is mostly around 10% in the other databases; a difference in definition or data collection method is likely to be the reason for this disparity. All databases place cycling rates at around 3%.

Cape Town

Before South Africa's first National Household Travel Survey of 2003 (NHTS, 2003), little was known about NMT travel behaviour in the country

(Behrens, 2009), and there was no regulatory requirement to collect such data (PGWC, 2009). The 2003 Survey indicated that walking, as the main mode, varied from 9% (high-income households) to 43% (middle-income households) and 61% (low-income households) (Behrens, 2009), and that cycling was still 'quantitatively insignificant', with the highest usage of bicycles (at 2%) recorded amongst poorer commuters (Jennings, 2015). Overall, the 2003 NHTS indicated that bicycles made up approximately 0.4% of modal split in Cape Town.

In 2013, the City conducted its own Integrated Public Transport Network Household Survey (CCT, 2013); this survey reported that the most popular main mode of travel was by car as a driver (25%), followed by walking (21%), minibus taxi (paratransit) (15%), private car passenger (12%) and train (11%). Walking has the highest percentage in the low-income group (33%).

In Cape Town, four reliable historic data sources were identified that provided modal splits for all trips. The percentages in the different databases range substantially (see Figure 2.1). For walking, the National Household Travel Survey (NHTS) 2013 identifies 20% as the mode share, while the Census 2011 found that 34% of people in Cape Town walk. For other modes, the different ranges are also substantial: public transport ranges between 26% and 38%, while travel by private vehicle ranges between 34% and 50%. All databases found the percentage of bicycle use low (between 0.4% and 1%).

ACET travel behaviour survey: the respondents

The aim during the data collection phase was to select similar households to maximize the comparison possibilities of cities. Table 2.3 presents various differences and similarities.

Cape Town appears to describe a proportion of population that has a higher average age than the other two case cities. Dar es Salaam has a household size that is substantially higher than for the other two case cities. Furthermore, Dar es Salaam has the highest proportion of teens (age group 13 to 17 years, who will, most likely, attend secondary school). Nairobi has the highest proportion of working people (age group 18 to 60 years), at 77.6%, while Dar es Salaam has 66% and Cape Town 63%.

ACET travel behaviour survey: modal split

All trips

When analyzing all trips (every trip segment in a trip is recorded separately) made in Cape Town, Dar es Salaam and Nairobi (see Figure 2.2), it is clear that NMT is the most important mode. Over 70% of people in Nairobi and Dar es Salaam use NMT modes, while a further 22.9% to 26.7% use (formal or informal) public transport, which will, most likely, include an NMT trip

Table 2.3 Comparison of sampled population by gender and age

Gender/age categories	Cape Town		Dar es Salaam		Nairobi	
	Number	Percent	Number	Percent	Number	Percent
Persons per household	4.0		8.1		4.8	
Gender of household members						
Children below 5 years	514	6.4	680	4.2	623	6.5
Male	3,483	43.3	7,595	46.8	4,589	47.8
Female	4,052	50.3	7,948	49.0	4,388	45.7
Total	8,049	100	16,223	100	9,600	100
Age of household members in years						
0–4	555	6.9	680	4.2	623	6.5
5–12	1,001	12.4	2,057	12.7	962	10.0
13–17	734	9.1	2,264	13.9	521	5.4
18–40	3,114	38.5	8,631	53.2	6,485	67.6
41–60	1,996	24.7	2,122	13.1	961	10.0
Above 60	669	8.3	349	2.1	48	0.5
Refused to answer	24	0.3	128	0.8	0	0.0
Total	8,093	100	16,231	100	9,600	100

Source: Masaoe et al., 2011.

segment to and from the stop or station. In Cape Town the percentages are lower (46.7% NMT and 23.2% public transport), but still significantly more than other modes. The difference in income and car ownership is reflected in this figure too, with almost 30% of respondents using private vehicles in Cape Town.

In the ACET travel behaviour surveys, walking accounts for 99% of all NMT trips in all three cities, while cycling makes up the other 1%.

Public transport services in Dar es Salaam and Nairobi rely heavily on paratransit (70%). In Dar es Salaam, bus travel is responsible for the other 30% of public transport trip segments. Train and other public transport services barely feature. In Nairobi, bus travel accounts for almost 21%, train travel is a

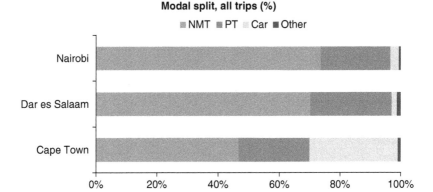

Figure 2.2 Modal split in Cape Town, Dar es Salaam and Nairobi (all trips in %).

Table 2.4 Public transport modes used in case cities for all trips (%)

PT mode	Cape Town	Dar es Salaam	Nairobi
Paratransit	57.2	70.0	70.0
Bus	14.7	29.7	20.9
Train	28.1	0.0	2.4
Other	0.0	0.3	6.7

Table 2.5 Car trips in case cities for all trips (%)

Car mode	Cape Town	Dar es Salaam	Nairobi
Driver	60.4	70.1	70.9
Passenger	39.6	29.9	29.1

mere 2.4%, and other public transport services (such as scooter taxi) account for 6.7% of the public transport trip segments in Nairobi.

Public transport in Cape Town is clearly different to the other two case cities. Paratransit accounts for 57.2%, while bus travel accounts for 14.7%. Train travel is an important public transport mode in Cape Town, and in the ACET database accounts for 28.1% of all trips.

Table 2.4 provides the split between public transport modes for all trips that users in the ACET database make.

Car users are either the driver or a passenger. Table 2.5 provides the split between the two for all trips in the ACET database.

In the three cities, the majority of car users are drivers (60% to 70%), while 29% to 40% of the car users are passengers. This amounts to an average occupancy of 1.66 in Cape Town and 1.42 in Dar es Salaam and Nairobi. In 2005, Vanderschuren collected data on one of the highways in Cape Town and found an identical value (Vanderschuren, 2006).

It is interesting that occupancies are so low in these three developing cities, especially taking all trips into account. In developing countries it is expected to see more carpooling, especially by poorer residents. This unexpected finding may be because the urban wealthy do not carpool with the urban poor. In the case of South Africa, the NHTS 2013 found that 65% of the population do not have access to a private vehicle (either within their household or the neighbourhood).

Furthermore, low occupancies are more common for the morning peak (see, for example, Gordon and Wong, 1985). As this data is for a 24-hour period, higher values were expected. It was, therefore, concluded that vehicle occupancy rates are, generally, low in the case cities.

Main mode

In many travel behaviour surveys, only the main mode of transport is used (i.e. NHTS databases, 2003 and 2013; also see Vasconcellos, 2013). In other words, only one trip segment of a trip is recorded (i.e. when a public transport user walks to the public transport stop, uses transit and then walks to the destination, the only recorded segment is the public transport segment).

For the three cities (see Figure 2.3), Nairobi has 61.3% of trips made by NMT only, while Dar es Salaam and Cape Town record 48.4% and 33.9%, respectively. When analyzing a 24-hour period, the percentage of trips that involved walking (or cycling) all the way is much higher than the percentages found in databases that focus on the morning peak, such as the NHTSs referenced in this chapter.

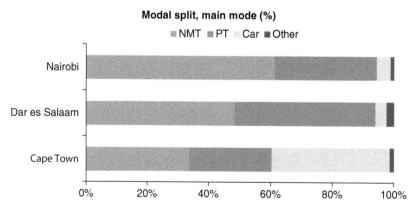

Modal split, main mode (%)

NMT ▪ PT ░ Car ▪ Other

Figure 2.3 Modal split for the case cities (main mode in %).

In both Nairobi and Dar es Salaam, 94% of trips are either by NMT or public transport, indicating that 94% of trips, either completely or partly, use NMT modes, as NMT is the most common mode to/from public transport. In Cape Town, the NMT and public transport combination accounts for just over 60% of trips, in total. Car trips account for 38.4% in Cape Town, while in the two other cities, the percentage is less than 5% for car travel. Given the large percentage of trips made solely or completely by NMT modes, the requirements for improved infrastructure provision are abundantly clear (also see Chapters 7 and 8).

Modal split to work

In many cases, researchers and governments are interested in the modal split to work (i.e. NHTS databases, 2003 and 2013; also see Vasconcellos, 2013). In Nairobi, 41.2% of people use NMT all the way to work, while 33.6% of people do so in Dar es Salaam; only 12.2% walk all the way to work in Cape Town. Public transport as the main mode accounts for 57.6% (Dar es Salaam), 51.2% (Nairobi) and 35% (Cape Town) of work trips. As indicated, the trip segments to/from public transport are mostly walking. In Nairobi and Dar es Salaam, NMT is, either completely or partly, used for the work trip for 92.4% and 91.2%, respectively. Even in Cape Town, where a large portion of people use their private vehicles (47.7%) to travel to work, the portion of trips with an NMT component (i.e. NMT used for the whole or part of the trip) is almost equal (47.2%).

NMT modes are undoubtedly the dominant mode in Nairobi and Dar es Salaam, and in Cape Town NMT is a significant mode at some 47% of trip segments (see Figure 2.4).

Trips to education

NMT is the dominant means of travel to an educational institution in all three cities (see Figure 2.5). In Nairobi, 68.9% of travellers, for educational

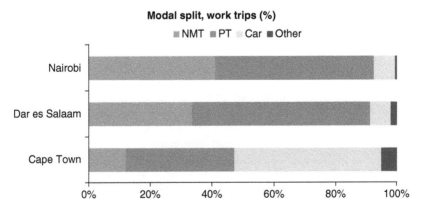

Figure 2.4 Modal split for work trips in case cities (main mode in %).

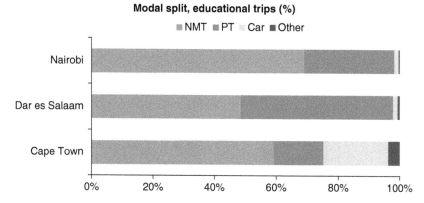

Figure 2.5 Modal split for educational trips in case cities (main mode in %).

purposes, use NMT as their only (main) mode of transport, while 59.2% and 48.6% use only NMT in Cape Town and Dar es Salaam, respectively. In Dar es Salaam, longer distances mean that 49.2% of scholar trips are by public transport. In Nairobi and Cape Town fewer travellers, for educational purposes, use public transport (29.3% and 15.9%, respectively). Again, these public transport trips will include walking (or cycling) to and from the public transport stop/rank/station.

Travel times and access to public transport

As trip segments to and from public transport stops and stations are mostly by means of walking (or cycling), access to public transport services is essential, and travel time to public transport is an important performance indicator (questions related to travel time rather than travel distance are more likely to receive reliable answers from travel survey respondents) (Svenson et al., 2012). Figure 2.6 provides an overview of the average travel times to the various public transport services in Cape Town, Dar es Salaam and Nairobi.

Paratransit services can be accessed in all three cities within around 10 minutes. Bus services, on the other hand, take between 11.2 (Cape Town) and 17.0 minutes (Nairobi), while train services take between 26.8 minutes (Cape Town) and 29.3 minutes (Nairobi), on average. Dar es Salaam has few (if any) train and bus services, as can be seen in Figure 2.6.

In the Millennium Development Goals (MDG; UN Millennium Project, 2005), improved access to schools and water was translated to a walking time of a maximum of 1 hour (per direction). Table 2.6 proves that, although the average walking time is less than 1 hour, there are NMT users that travel longer than 1 hour.

Taking all trip segments into account, average walking travel time in the three case cities varies between 12 minutes in Cape Town to 19 minutes in

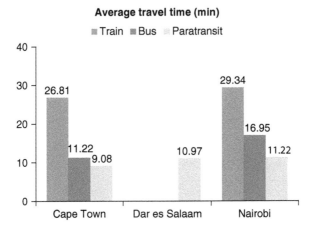

Figure 2.6 Average travel time in case cities (minutes).

Table 2.6 Walking travel times in case cities (all trip segments)

Travel time	Cape Town	Dar es Salaam	Nairobi
Average (min)	12	19	15
Minimum (min)	0.5	2	01
Maximum (hh:mm)	4:00	3:14	2:34
Standard deviation (min)	19	19	15

Dar es Salaam. The minimum travel time is less than one minute in Cape Town and two minutes in Dar es Salaam. Maximum walking travel times are the least in Nairobi (2:34 hours), while Cape Town has walking times of up to four hours. Finally, the standard deviation of walking travel times in the three case cities are substantial, i.e. at least as much as the average travel time.

Personal characteristics of NMT users: analysis by gender, age and income

During the process of data collection by ACET, all household members were asked to complete a trip diary. Gender and age distributions are, therefore, not influenced by the data collection method and are a true reflection of the household members interviewed. The overall NMT gender split is provided in Figure 2.7.

For walking, the gender split is fairly even. In Cape Town, women make up over 56% of pedestrians. In Dar es Salaam, the percentage of women

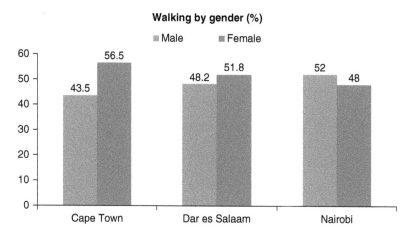

Figure 2.7 Walking by gender in case cities (%).

who walk is also slightly higher (almost 52%) than the percentage of males. On average, males tend to be more active in the economy and, therefore, use relatively more motorized modes (see, for example, STATSSA, 2012). In Nairobi, more men (52%) than women seem to walk.

The gender split for cycling is very different. The majority of cyclists are male. In Cape Town and Dar es Salaam the percentage of male cyclists is just over 81%, while in Nairobi almost 91% of cyclists are male. Cycling is clearly not as attractive to women. The literature lists various reasons for the low percentage of women cycling; the most common are the security risks (see, for example, Garrard et al., 2012) and cultural barriers (Pochet and Cusset, 1999).

Regarding the analysis of NMT use by age, children below the age of 12 generally travel to primary school; between the ages of 12 and 18, learners travel to high school, and after age 18, work trips become a possibility. A driver's licence is an option at the age of 18. The working ages, between 18 years and 60 years, were split in two, to add more detail (see Figure 2.8).

The majority of pedestrians are between 19 years and 45 years old (working age). In Cape Town, the portion of NMT users in this category is 46.5%, while Dar es Salaam and Nairobi have higher percentages: 57.9% and 76.5%, respectively. Trips were counted as follows: if a person walked 100% of the way, this is counted as one trip; if a person walked to the bus, travelled further, then walked to their final destination, this is counted as one bus trip and two walking trips. This is why the main mode of transport is used for mode analysis although this research shows that, in many instances, NMT trips are not counted at all.

As indicated earlier, the travel diaries were recorded for each member of a household and the data collection method does not influence the age

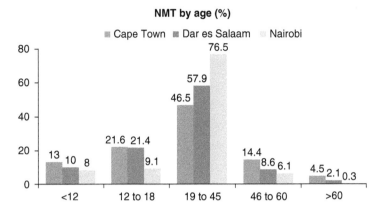

Figure 2.8 NMT users by age in case cities (all trip segments in %).

distribution. Furthermore, the population distribution in Africa includes a large percentage of young people, i.e. almost 40% of people living in Africa are under the age of 15 years old (www.geohive.com). It can, therefore, be concluded that the 19- to 45-year-old age group is more economically active than the other age groups, even if the difference in age group size is taken into account. There is a clear drop in NMT trips for the age group between 46–60 years old. This can be explained by the reduced population numbers (only 11.4% of Africans fall within this age group; www.geohive.com).

In Dar es Salaam the lowest income groups use NMT the most. There is an exponential drop of NMT use as income levels increase, from just over 45% of people with an income of less than US$134 using NMT (mostly walking) to 6% or less from an income level of $401 or more. In Nairobi the exponential drop is witnessed from an income level of US$201. However, for the first three income categories, the use of NMT is steady (between 22% and 24%).

In Cape Town the income distribution of NMT users is very different. The first two income categories (less than US$134 and US$135–US$200) is low, i.e. 8.4% and 6.2%, respectively. This is probably due to higher income levels in Cape Town. From US$201, all categories show a significant portion of NMT users. The willingness of people with higher incomes to use NMT is clearly higher in Cape Town. In every income group, at least 2.8% of trip segments are by NMT. Furthermore, in the highest income group, 6.8% of trips are by NMT.

Conclusion

What are the implications of the ACET travel behaviour survey findings for NMT policy and practices in Sub-Saharan African cities? In 2001, Rwebangira, reporting for the World Bank, cautioned that the existing transport policy and planning framework inadequately served the needs of people walking and cycling, through providing infrastructure and addressing road safety problems. The findings of this travel survey indicate that 25 years after the above-cited work, NMT modes remain the dominant modes. Later chapters argue that, despite many decades of concern regarding the poor provision of infrastructure for this major mode, and the road safety records (see Chapters 4–6), political attention, revenue allocation and facilities provision has not yet sufficiently shifted to NMT users (see Chapters 7, 8, 12 and 13). NMT users continue to bear the inequitable burden of road-crash-related injuries and fatalities.

References

Behrens, R. 2009. What the NHTS reveals about non-motorised transport in the RSA. In Khadpekar, N. (ed), *Non-motorised transportation: making it a viable option.* Ahmedabad: ICFAI University Press.

Behrens, R. and Masaoe, E. 2009. Comparative experimental application of alternative travel diaries in Cape Town and Dar es Salaam, 28th Southern African Transport Conference: Sustainable transport, Pretoria.

City of Cape Town (CCT). 2012. Statistics for the City of Cape Town – 2012. Available at: http://resource.capetown.gov.za/documentcentre/Documents/Maps%20and%20statistics/City_Statistics_2012.pdf

City of Cape Town (CCT). 2013. Development of a city wide integrated public transport network. Household survey report.

Garrard, J., Handy, S. and Dill, J. 2012. Women and cycling. In Pucher, J. and Buehler, R. (eds), *City cycling.* Cambridge, MA and London: MIT Press.

Gordon, P. and Wong, H.L. 1985. The costs of urban sprawl: some new evidence. *Environment and Planning A,* 17: 661–666.

Howe, J. and Bryceson, D. 2000. *Poverty and urban transport in East Africa: review of research and Dutch Donor experience.* Washington, D.C.: World Bank.

Howe, J. and Dennis, R. 1993. *The bicycle in Africa – luxury or necessity?* IHE Working Paper IP-3. Delft, The Netherlands presented at VELOCITY Conference 'The Civilised City': Responses to New Transport Priorities, Nottingham, UK, 6–10 September.

Jennings, G. 2015. A bicycle revolution in SA: A critique of trends, policies and programmes. Southern African Transport Conference, Pretoria, July 2015.

Kenya National Bureau of Statistics (KNBS). 2010. Database. Available at: www.knbs.or.ke/

Khayesi, M., Monheim, H. and Nebe, J.M. 2010. Negotiating 'streets for all' in urban transport planning: The case for pedestrians, cyclists and street vendors in Nairobi, Kenya. *Antipode,* 42(1): 103–126.

Masaoe, E., des Mistro, R. and Makajuma, G. 2011. Travel behaviour in Cape Town, Dar es Salaam and Nairobi Cities. South African Transport Conference, July 2011.

National Household Travel Survey (NHTS). 2003. South African Wide Database. Pretoria, South Africa: Statistics South Africa.

National Household Travel Survey (NHTS). 2013. South African Wide Database. Pretoria, South Africa: Statistics South Africa.

Pendakur, V.S. 2005. Non-motorized transport in African cities: lessons from experience in Kenya and Tanzania. *SSATP Working Paper No. 80.* Washington, D.C.: World Bank.

Pochet, P. and Cusset, J.-M. 1999. Cultural barriers to bicycle use in Western African cities. *IATSS Research,* 23(2): 43–50.

Provincial Government Western Cape (PGWC). 2009. NMT in the Western Cape Draft Strategy. Cape Town, South Africa.

Rwebangira, T. 2001. Cycling in African cities: status and prospects. *World Transport Policy and Practice*, 7(2): 7–10.

Servaas, M. 2000, December. The significance of NMT for developing countries. Washington, D.C.: World Bank.

South African Government Gazette. 2009. No. 5 of 2009: National Land Transport Act, Vol. 526. Cape Town, 8 April 2009, No. 32110.

South Africa Bureau of Statistics (STATSSA). 2011. South African Census 2011. Available at: www.statssa.gov.za/?page_id=3839.

Statistics South Africa (STATSSA). 2012. *Gender statistics in South Africa – 2011*. Pretoria, South Africa. Available at: www.statssa.gov.za/publications/Report-03-10-05/Report-03-10-052011.pdf.

Sub-Saharan Africa Transport Policy Program (SSATP). 2015. Policies for sustainable accessibility and mobility in urban areas of Africa. Working Paper No. 106, June 2015.

Svenson, O., Eriksson, G. and Gonzalez, N. 2012. Braking from different speeds: judgments of collision speed if a car does not stop in time. *Accident Analysis and Prevention*, 45: 487–492.

Tembele, R. 2000. Productive and liveable cities. Guidelines for pedestrian and bicycle traffic in African cities. Amsterdam: Tanzanian National Team for Non-Motorised Transport Vélo Mondial 2000.

UN Millennium Project 2005. Investing in development: A practical plan to achieve the Millennium Development Goals. Overview. London: Earthscan, ISBN: 1-84407-217-7.

United Republic of Tanzania (URot). 2013. Population distribution by administrative units, key findings, 2012 population and housing census. Dar es Salaam: National Bureau of Statistics.

Vanderschuren, M.J.W.A. 2006. Intelligent transport systems in South Africa. Impact assessment through microscopic simulation in the South African context. TRAIL Thesis Series T2006/4, ISBN 9055840777, August 2006.

Vasconcellos, E.A. 2013. Urban transport environment and equity: the case for developing countries. New York: Earthscan (first published in 2001).

3 Pedestrian crossing behaviour in Cape Town and Nairobi

Observations and implications

Roger Behrens and George Makajuma

Introduction

Sub-Saharan African cities experience high road crash fatality rates. In Cape Town and Nairobi – the case cities examined in this chapter – pedestrians account for as much as 60% and 80% of road crash fatalities respectively (Behrens, 2005; Mengot, 2012). Pedestrians crossing the road without the use of crossing facilities are the greatest cause of fatal crashes, accounting for around 36% of the factors contributing to all fatal crashes in South Africa (Behrens, 2002). Data on the relationship between pedestrian road crash fatalities and road classification are sparse and outdated. Those that are available (CoCT, 2004; Ribbens, 1990) indicate that, because of the greater speed differential, fatal pedestrian crashes occur mostly on arterials and freeways. Speed differential is central to understanding crash risk: the probability of a fatality resulting from a crash involving a vehicle and a pedestrian increases exponentially with speed at collision. The relationship with collision speed follows a sigmoidal curve in which fatality probability begins to increase rapidly after 40 kilometres/hour, and levels off at around 90 kilometres/hour where the probability of fatality becomes fairly certain (Rosén et al., 2011).

To address the pedestrian safety problem in Cape Town and Nairobi, it is therefore essential that pedestrian crossing behaviour and attitudes, as well as pedestrian–driver interaction, along arterials and freeways are understood. While pedestrian road crossing behaviour has been the subject of extensive research elsewhere (see, for instance, Ishaque and Noland, 2008 for a comprehensive review of literature in this field, and Papadimitriou et al., 2009 for a review of alternative approaches to pedestrian modelling), to date, little research of this nature has been carried out within the Cape Town or Nairobi context. The bulk of the international research has focused on the walking speed of different types of pedestrians while crossing different types of crossing facilities, in order to better understand crossing delay, gap acceptance and signal phasing requirements, and the reasons for temporal and spatial non-compliance with crossing regulations. Of significance to the research reported upon in this chapter, studies of spatial non-compliance (e.g. Chu et al., 2004; Sisiopiku and Akin, 2003) have found that the extra walking distance required

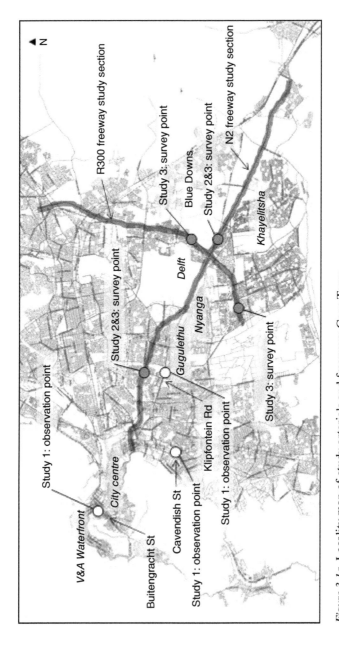

Figure 3.1a Locality map of study arterials and freeways: Cape Town

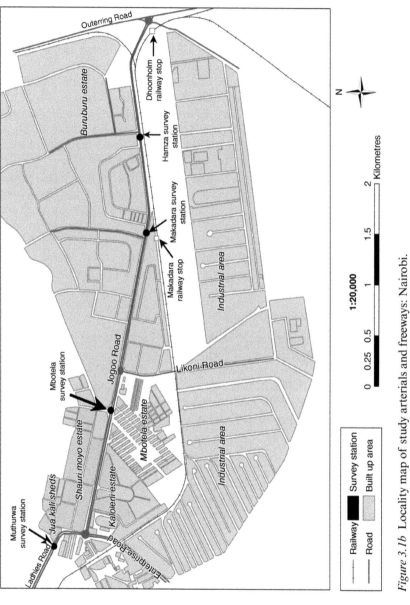

Figure 3.1b Locality map of study arterials and freeways: Nairobi.

to reach a crossing facility is an important contributing factor in the decision to jaywalk, although the likelihood of spatial non-compliance decreases as traffic volumes increase, which suggests an association between levels of non-compliance and risk. A limitation of this body of research, from the perspective of understanding pedestrian crossing behaviour in Cape Town and Nairobi, is that it focuses largely on compliance with traffic control systems at at-grade pedestrian crossing facilities that regulate the location and time of crossing. Little research appears to have been undertaken on illegal at-grade freeway crossing behaviour. As will be demonstrated later in this chapter, due to a combination of high walking dependency and weak law enforcement, this behaviour is common in the South African and Kenyan contexts.

The purpose of this chapter is to report upon a series of small indicative studies, undertaken between 2004 and 2010, that observed pedestrian crossing behaviour and collected attitudinal information on selected arterials and freeways in Cape Town and Nairobi, using a variety of methods.

The chapter is divided into four sections. In the following section, two studies of arterial crossing – the first in Nairobi and the second in Cape Town – are discussed in terms of their research methods and key findings. In the third section, two studies of freeway crossing – both conducted in Cape Town – are discussed in terms of their methods and findings. The chapter concludes, in the final section, with a synthesis of findings, and a discussion on implications for developing planning and management practices that improve pedestrian safety.

Arterial crossing behaviour studies

The arterial crossing study in Nairobi sought to understand the dynamics of pedestrian crossing behaviour. It focused on Jogoo Road, an arterial along which travel conditions for pedestrians are severely degraded as a result of inequitable provision of infrastructure for all categories of road users.[1] The Jogoo Road corridor is one of the major arterials linking Nairobi's central business district (CBD) to the populous low-income Eastlands area of the city, which stretches as far as the Outer Ring Road (see Figure 3.1b). The study section is 5.6 kilometres long. There are two roundabouts along the road: one at the City Stadium end where Jogoo Road meets Landhies Road; and one at the end of the Jogoo Road link towards the Jomo Kenyatta Airport. The road is basically straight and experiences high pedestrian activity, which makes it an ideal study site. It connects the middle to low-income residential estates of Buruburu, Umoja, Donholm and Embakassi, extending up to Jomo Kenyatta International Airport. There also exist pockets of informal settlement lying along the periphery of these middle-income estates. The low-income housing areas are the main generators of pedestrian trips. Most residents earn their living from the informal sector (*jua kali* industry) found predominantly in this part of the city. Trip patterns are normally diffused: dispersed both in space and time.

The arterial crossing study in Cape Town sought to test a starting hypothesis, or supposition, that the plotted distance of observed unassisted pedestrian crossing points from the nearest crossing facility would follow a sigmoidal curve. Closer to the crossing facility, fewer unassisted pedestrian crossings were expected than further away, where detour refusal rates[2] were anticipated to be higher. It was posited that the start of the sigmoidal curve would differ in accordance with variations in perceived vehicle collision risk and perceptions of the likelihood of severe casualty. Thus it was anticipated that the curves of arterials and freeways would start further from the crossing facility than the curves of collector routes on which crossing facilities are warranted, and that, in turn, the curve of a freeway would start further from the crossing facility than the curve of an arterial. Figure 3.2 illustrates this hypothesis diagrammatically.[3] It was believed that understanding the characteristics of this curve, and how it differed across road classes, would add valuable insights into the spacing between crossing facilities required to reduce unassisted and illegal pedestrian crossing behaviour and thereby improve safety. More specifically, if the distance values of (x) and (y) in Figure 3.2 could be derived from empirical observations, a doubling of this value would present the spacing interval of crossing facilities, along arterials and freeways respectively, that matched the willingness of pedestrians to detour from their desire line[4] in order to cross at a safer point.

The study, undertaken in the spring of 2004, observed pedestrian crossing points on two arterials (Klipfontein Road and Buitengracht Street) and on one major collector (Cavendish Street) (Naidoo, 2004). The selected section of Klipfontein Road (between the intersection with Vanguard Drive and Hazel Road) is a dual carriageway with three traffic lanes in each direction. A median separates the opposing lanes. Signalized at-grade crossing facilities are provided at the road intersections and in the midblock. Directional peak hour traffic volumes are in the region of 1,350 vehicles/hour and the posted speed limit is

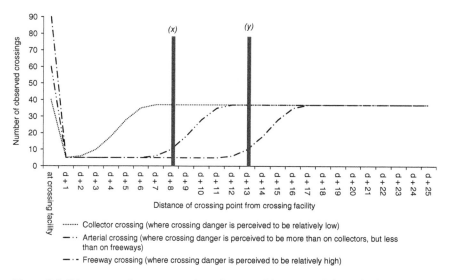

Figure 3.2 Diagrammatic representation of overarching research hypothesis.

70 kilometres/hour. The selected section of Buitengracht Street (between the intersection with Coen Steytler Avenue and Hans Strijdom Avenue) is a dual carriageway with five and six traffic lanes in opposing directions. A wide median separates the opposing lanes. Signalized at-grade crossing facilities are provided at the road intersections. Peak hour traffic volumes are in the region of 2,500–3,500 vehicles/hour/direction and the posted speed limit is 60 kilometres/hour. Cavendish Street (between Osborne Road and Vineyard Road) is a dual carriageway with two traffic lanes in each direction. Narrow pedestrian refuge islands separate the opposing lanes. An unsignalized at-grade crossing facility is provided in the midblock. Directional peak hour traffic volumes are in the region of 1,500 vehicles/hour and the posted speed limit is 60 kilometres/hour.

Research method

Study 1: observed arterial crossing behaviour in Nairobi

The arterial crossing study in Nairobi employed manual count techniques and (n = 820) pedestrian crossing observations at four selected sites. Pedestrian crossings were observed during off-peak periods as this is when vehicle speeds are high and conditions are perceived to be most dangerous for pedestrians and bicyclists. The survey points were selected to correspond with the predominant pathways[5] of pedestrians on the corridor. The crossings were timed, and pedestrian movement characteristics were noted in terms of crossing trajectories, how relaxed pedestrians appeared when crossing the roadway, whether crossing was undertaken in a group, and whether crossing was undertaken while running. Two fieldworkers manned each crossing location, one recording the nature of movement across the roadway while the other selected a pedestrian randomly and timed the waiting and crossing times. A pedestrian was selected randomly after a one-minute time lapse of a complete timed crossing. The first batch of data was collected in the month of January 2009, and additional traffic speed and volume data were collected later in the month of November 2010 in order to assess the impact of vehicle speed and volume on pedestrian crossing behaviour. The survey points are indicated in Figure 3.1b.

Study 2: observed arterial crossing behaviour in Cape Town

Two methods of observation were utilized in the arterial crossing study in Cape Town. The first method (in the case of Buitengracht Street and Klipfontein Road) took the form of analysis of recorded crime surveillance video footage obtained from the City of Cape Town. Footage from weekday morning and evening peak periods were analysed (15h45–17h45 on Klipfontein Road, and 06h00-07h15 and 16h00-17h45 on Buitengracht Street). Observed pedestrian crossing points were recorded on street plans. In order to estimate the distance of crossing points from crossing facilities accurately, site inspections were conducted in which regularly spaced landmarks were identified and their distance from the nearest crossing facility measured. The limited visual range of the video cameras placed limitations on the length of the road sections observed. The second method (in the case of Cavendish Street) took the form of manual roadside counts and measurements. These observations were conducted on a Saturday morning during the peak shopping period (11h00–12h00). Fieldworkers were stationed between two intersections, and marked observed pedestrian crossing movements on a street plan. The limited visual range of the fieldworkers also placed limitations on the length of the road section observed.

Research findings

Study 1: observed arterial crossing behaviour in Nairobi

From the tabulated crossing times in Table 3.1, there is evidence from the Nairobi study that it takes relatively long for pedestrians to find adequate gaps in the traffic stream to cross the roadway (the mean waiting time in the

Table 3.1 Pedestrian roadway crossing time, by location (n = 820)

Crossing location		Min. (sec)	Max. (sec)	Mean (sec)	Standard deviation
Muthurwa (n = 210)	Time taken to cross first half of the road	1	8	4.9	1.2
	Waiting time at the median	1	15	3.9	1.6
	Time taken to cross second half of the roadway	1	15	4.4	2.2
Mbotela Stage (n = 200)	Time taken to cross first half of the road	1	10	5.1	1.7
	Waiting time at the median	1	38	15.3	9.9
	Time taken to cross second half of the roadway	0	11	5.6	1.7
Makadara Footbridge (n = 200)	Time taken to cross first half of the road	1	19	4.6	2.3
	Waiting time at the median	1	39	13.7	8.7
	Time taken to cross second half of the roadway	2	19	5.1	2.3

(continued)

Table 3.1 Pedestrian roadway crossing time, by location (n = 820) (*continued*)

Crossing location		Min. (sec)	Max. (sec)	Mean (sec)	Standard deviation
Hamza Stage (n = 210)	Time taken to cross first half of the road	1	20	6.1	2.6
	Waiting time at the median	2	99	16.2	13.0
	Time taken to cross second half of the roadway	3	66	8.4	6.7

median was 12 seconds, and maximum 99 seconds). At busy sections like the Hamza crossing point, the crossing manoeuvres take longer to be completed as pedestrians get trapped and sandwiched between the two traffic streams as a result of overcrowding on the roadway by vehicles. This leads to frustration and anxiety amongst crossing pedestrians, and as a consequence, many run across the roadway with associated vehicle collision risks.

Further, considering a typical walking speed of 1.1 metres/second for pedestrians,[6] the minimum time needed to comfortably (safely) cross a 7-metre carriageway would be 7.7 seconds for a two-lane roadway (De Langen and Tembele, 2001). The Jogoo Road data indicate that, on average (excluding wait times), pedestrians take 5.5 seconds to cross the 7-metre-wide carriageway (at 0.8 metres/second), with the recorded minimum time of 4.6 seconds at the Mbotela crossing point (at 0.7 metres/second). These data indicate that during the off-peak, pedestrians generally run across the road in response to inadequate gaps for crossing at typical walking speeds.

The difficulties experienced by crossing pedestrians are further depicted by crossing patterns in Table 3.2. Pedestrians seldom cross at the crossing facilities provided because motorists rarely stop to allow them to cross (82% of observed crossings were away from designated crossing points). The crossing behaviour presented in Table 3.3 indicates a strong perception of crash risk by pedestrians: at both the Mbotela and Makadara crossing points, significant number of pedestrians cross the roadway in a diagonal manner, as they continue looking for adequate gaps for crossing away from zebra crossings.

Pedestrians often bunch together to cross a roadway. This is thought to impose a psychological challenge to force speeding motorists to slow down or yield. This effect is often observed on roads with narrow operating speed ranges which border the operating speed limits in urban roadways, as was observed along Landhies Road at the Muthurwa[7] crossing where the minimum spot speed observed was 6.4 kilometres/hour and the maximum speed was 52.4 kilometres/hour.[8] At such speeds, the bunching effect of pedestrians is able to impose a psychological effect on drivers, and enables safer crossings

Table 3.2 Pedestrian crossing points in relation to crossing facilities (n = 820)

Crossing location		Frequency	Percent
Muthurwa (n = 210)	Besides (within 20m)	149	71.0
	Besides (20–100m from crossing point)	61	29.0
Mbotela Stage (n = 200)	Crossing point (RZC/painted)	95	47.5
	Besides (within 20m)	70	35.0
	Besides (20–100m from crossing point)	35	17.5
Makadara Footbridge (n = 200)	Crossing point (RZC/painted)	54	27.0
	Besides (within 20m)	105	52.5
	Besides (20–100m from crossing point)	41	20.5
Hamza Stage (n = 210)	Besides (20–100m from crossing point)	210	100.0

Table 3.3 Pedestrian crossing trajectory (n = 820)

Crossing location		Frequency	Percent
Muthurwa (n = 210)	Crossed at right angle	138	65.7
	Crossed diagonally	72	34.3
Mbotela Stage (n = 200)	Crossed at right angle	94	47.0
	Crossed diagonally	106	53.0
Makadara Footbridge (n = 200)	Crossed at right angle	121	60.5
	Crossed diagonally	79	39.5
Hamza Stage (n = 210)	Crossed at right angle	101	48.1
	Crossed diagonally	109	51.9

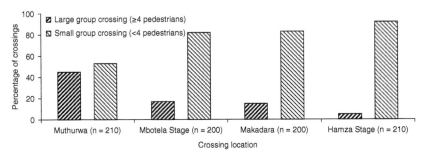

Figure 3.3 Size of pedestrian crossing group (n = 820).

Table 3.4 Vehicle spot speeds (kilometres/hour)

| | Survey station | | | | | | | |
| | Donholm to CBD *(inbound)* | | | | CBD to Donholm *(outbound)* | | | |
	Hamza	Landhies	Makadara	Rikana	Hamza	Landhies	Makadara	Rikana
Mean speed	58.9	28.2	30.7	27.9	48.1	31.9	35.3	33.3
Min. speed	14.1	7.1	2.7	1.0	14.7	6.4	12.0	6.6
Max. speed	114.9	52.4	94.8	45.4	128.6	52.4	72.0	52.4
Median speed	58.7	28.3	29.8	29.8	45.8	32.0	36.0	34.4

of roadways. The bunching effect was, however, less effective at sections on the corridor with less uniform traffic speeds characterized by wide traffic stream speed differentials (see Figure 3.3). On Jogoo Road, particularly at the Hamza crossing, the speed survey indicated a speed range of 14.7 to 128.6 kilometres/ hour in the outbound direction (see Table 3.4). Such extreme fluctuations in vehicle operating speeds often depict unpredictability in motorist behaviour, which raises the level of crash risk as perceived by pedestrians.[9] The 'safety-in-numbers' effect arising from the bunching of pedestrians does not seem to work therefore under wide driver speed variability: crossing manoeuvres are then taken on an individual basis. At higher varying speeds, pedestrians are generally intimidated and seldom attempt to manipulate traffic in this way.

Study 2: observed arterial crossing behaviour in Cape Town

With respect to the Cape Town study of spatial compliance, of the pedestrians observed crossing the selected section of Klipfontein Road during the two hour observation period, only 15% of crossings were observed to occur at the crossing facility. The remaining 85% of crossings were at other points. Figure 3.4a

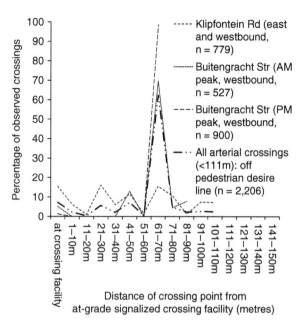

Figure 3.4a Observed pedestrian crossing distance from at-grade signalized facility, by arterial study (Naidoo, 2004): crossing facility located off pedestrian desire line (n = 2,206).

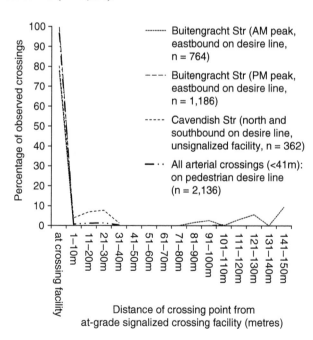

Figure 3.4b Observed pedestrian crossing distance from at-grade signalized facility, by arterial study (Naidoo, 2004): crossing facility located on pedestrian desire line (n = 2,312).

illustrates the distribution of crossing points away from the crossing facility.[10] No particular pattern of crossing or pedestrian desire line could be established, and it was found that many pedestrians refused to detour even a short distance to utilize a crossing facility.

A very different pattern of crossing was observed on Buitengracht Street however. In the case of crossing facilities located away from the dominant pedestrian desire line (associated with workers moving between the central rail station and the Victoria and Alfred Waterfront development), 1–5% of crossings were observed at the crossing facility on the westbound lanes. The remaining 95–99% of crossings was distributed elsewhere, with a high concentration at 61–70 metres from the crossing facility at the point of intersection with the dominant pedestrian desire line. Figure 3.4a illustrates the distribution of crossings on the westbound lanes. In the case of crossing facilities located on the dominant pedestrian desire line (the eastbound lanes), 78–99% of pedestrian crossings were observed on the crossing facility. Figure 3.4b illustrates the distribution of crossings on the eastbound lanes. These observations suggest that the use of crossing facilities is closely linked to their location in relation to pedestrian movement desire lines.[11]

To explore the relationship between crossing facility use and pedestrian movement desire lines further, a case was sought where the crossing facility was located at the entrance of a large pedestrian trip attractor with no other entrance along that side of the street. Cavendish Street was selected for this purpose as it is characterized by an unsignalized crossing facility located directly at the entrance of the Cavendish Square shopping centre. Figure 3.4b illustrates the distribution of crossings on Cavendish Street. Consistent with the findings on eastbound Buitengracht Street lanes, 80% of crossings were observed at the crossing facility.

Freeway crossing behaviour studies

The initial intention of the Cape Town arterial study discussed above was to include observations of both arterial and freeway crossings. The freeway element of the study was, however, abandoned because suitable video footage could not be obtained and roadside observations presented security and visual range problems. The need for freeway crossing observation was, therefore, deferred to a later study (Mngomezulu, 2007), undertaken in the spring of 2007, which imputed and surveyed pedestrian crossing behaviour on two freeways (the N2 and the R300). The intra-city length of the N2 freeway extends 38 kilometres east-west, passing by a number of low-income residential neighbourhoods and informal settlements in the south-east of the city (including Gugulethu, Nyanga and Khayelitsha). Opposing traffic lanes are separated by a raised concrete barrier or median. The intra-city length of the R300 freeway extends 22 kilometres north-south, passing by a number of low-income neighbourhoods in the east of the city (including Delft and Blue Downs). Opposing traffic lanes are separated by a median. The maximum posted speed limit on both freeways is 120 kilometres/hour.

Given the insights that emerged from the small-sample intercept survey undertaken by Mngomezulu (2007), a further study, also conducted in Cape Town, was initiated to survey a larger, but nevertheless still indicative, sample of respondents which included both pedestrians who crossed at-grade and on a grade-separated facility. This study, undertaken in 2008, observed and surveyed pedestrian behaviour on the same two freeways (N2 and R300) (Gabuza, 2008).

Research method

Study 3: imputed and surveyed freeway crossing behaviour in Cape Town

The method of observation in the first freeway crossing study in Cape Town took the form of informal footpath recognition within freeway reserves from aerial photographs taken in 2005 (see Figure 3.1a for the extent of the N2 and R300 study sections), and verification by windshield observation of foot-paths and break-throughs in the concrete balustrade fencing separating the freeway reserve from bordering residential neighbourhoods. The distances between footpath crossings and the nearest grade-separated crossing facility (in the form of either a footway on a road bridge, or a footbridge) were meas-ured. In addition, in order to explore the attitudes of pedestrians crossing at-grade, and why crossing facilities were not utilized, an exploratory (n = 100) roadside intercept survey was conducted on two selected sections of the N2 freeway (115 metres west of the Vanguard Drive footway crossing facility, and 645 metres west of first footbridge east of the Swartklip Interchange). The questionnaire included 30 questions relating to the demographics of the respondent, the nature of the trip, reasons for crossing at that point, and attitudes towards crossing safety. Data were collected on weekdays between 14h00 and 18h00. Respondents were intercepted after the at-grade crossing movement.

Study 4: observed and surveyed freeway crossing behaviour in Cape Town

The method of observation in the second freeway crossing study in Cape Town took the form of a (n = 650) roadside intercept survey conducted on four selected sections at 10 intercept points (on the N2 freeway: 370 metres west and east of, and at, the Vanguard Drive footway crossing facil-ity, and 645 metres west of, and at, the first footbridge east of the Swartklip Interchange; and on the R300 freeway: 200 metres south of, and at, the first footbridge north of the Swartklip Interchange, and 900 metres west of, and at, the New Eisleben Drive footway crossing facility). These survey points were selected on the basis that they were identified as sites of high pedestrian at-grade crossing in the earlier aerial photograph analysis. The questionnaire included 20 questions relating to the demographics of the respondent, the nature of the trip, reasons for crossing at that point, and attitudes towards crossing safety. Data were collected on weekdays during morning and afternoon peak periods. Respondents were intercepted after the grade-separated or at-grade crossing movement.

Research findings

Study 3: imputed and surveyed freeway crossing behaviour in Cape Town

Aerial photography analysis of the N2 freeway in the first study revealed a total of 26 grade-separated crossing facilities (comprised of 4 footbridges, 2 subways and 20 footways on road bridges). Distances between crossing facilities vary, with a mean interval of 1,144 metres (and standard deviation of 1,101 metres). The minimum distance between crossing facilities is 234 metres, and the maximum is 4,343 metres. A total of 244 informal crossing footpaths were identified and verified. Figure 3.5 illustrates the distribution of footpath crossing distances from the nearest crossing facility. These findings indicate that, as in the case of arterials, some pedestrians are unwilling to detour even small distances to cross at a facility. The highest concentration of crossing footpaths was observed between 100 metres and 300 metres from the nearest facility. The decline of crossing footpaths beyond 700 metres can be attributed to crossing facility spacings, rather than to any behavioural pattern. The mean interval between crossing footpaths was found to be 123 metres, and the minimum and maximum intervals were found to be 10 metres and 741 metres

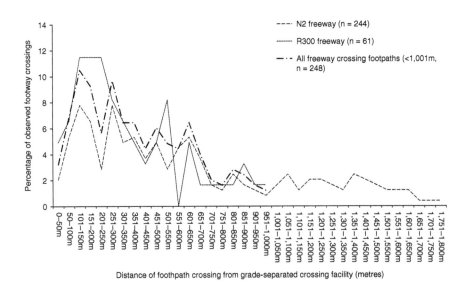

Figure 3.5 Imputed pedestrian crossing distance from grade-separated facility, by freeway study (n = 305; Mngomezulu, 2007).

respectively. The nearest crossing footpath to a crossing facility was found to be 33 metres, and the furthest 467 metres.

Aerial photography analysis of the R300 freeway revealed a total of 23 grade-separated crossing facilities. Distances between crossing facilities vary, with a mean interval of 961 metres (and standard deviation of 609 metres). The minimum distance between crossing facilities was 320 metres, and the maximum was 3,107 metres. A total of 61 informal crossing footpaths were identified and verified. Figure 3.5 illustrates the distribution of footpath crossing distances from the nearest crossing facility. These findings are broadly consistent with the N2 findings. The highest concentration of crossing footpaths was observed between 100 metres and 300 metres from the nearest facility. The mean interval between crossing footpaths was found to be 173 metres, and the minimum and maximum intervals were found to be 9 metres and 948 metres, respectively. The nearest crossing footpath to a crossing facility was 28 metres.[12]

Analysis of the intercept survey data revealed a disproportionate number of males in the sample (63% males vs. 37% females), and a greater number of younger people (40% of respondents were less than 20 years old, and 68% less than 25 years old). Most respondents were learners (34%), followed by workers (30%), unemployed persons (28%) and tertiary education students (8%).

When questioned on why they were crossing at that particular at-grade point, three single or combined most important reasons were given. Table 3.5 presents these findings, indicating that the most common reason was the desire to walk the shortest route (70%), followed by an equal concern for route distance and crime (17%) and then safety from crime (12%). The latter reason is associated with reports of criminals preying upon pedestrians trapped on footbridges

Table 3.5 Intercept survey respondents' stated reason for crossing at-grade, by perceived distance to the nearest crossing facility (n = 100; Mngomezulu, 2007)

Reason for crossing at chosen crossing point	Perceived distance to the nearest crossing facility (%)					Total
	0–100m	101–200m	201–300m	301–400m	> 400m	
Shortest route	18.0	18.0	8.0	16.0	10.0	70.0
Shortest route and safety from crime	3.0	5.0	5.0	1.0	3.0	17.0
Safety from crime	3.0	3.0	3.0	1.0	2.0	12.0
Item non-response						1.0
Total	24.0	26.0	16.0	18.0	15.0	100.0

without an escape route. In another question, 72% of respondents said that, for security reasons, they only use a crossing facility when they are not travelling alone.

When questioned on the three most important factors taken into account when selecting an appropriate crossing location more generally, a different response pattern emerged however. Table 3.6 presents these findings, indicating that safety from crime assumed greater importance in these responses, and particularly so amongst women. While this was a small sample and the results are therefore only indicative in nature, what is startling is the low or absent concern for traffic safety in the responses provided (with traffic volume and safety combined accounting for just 5% of responses). Nevertheless, 73% of respondents indicated elsewhere in the interview that they use grade-separated crossing facilities either when traffic volumes are high, at night, or in bad weather conditions. Twenty-seven percent of respondents reported that they never use grade-separated crossing facilities.

When questioned on the most important cause of pedestrian crashes on freeways, 80% of respondents indicated this was drunken pedestrians, followed by drunken motorists (15%), and then poor visibility (5%). Most respondents (97%) knew it is illegal to cross a freeway at-grade, and 16% reported that they knew a family member or a friend who had been involved in a pedestrian–vehicle collision.

Table 3.7 presents findings with respect to the relationship between grade-separated crossing facility use and the number of years the respondent had lived in a city. These data suggest that more frequent use of crossing facilities might be correlated with greater experience of urban life, and by implication, greater exposure to traffic risks. Amongst respondents who had lived in a city for five years or less, 54% always cross at-grade, compared to 9% of respondents who had lived in a city for greater than 20 years.

Table 3.6 Intercept survey respondents' stated important factors that influence freeway crossing decisions, by gender (n = 100; Mngomezulu 2007)

Important factors that influence freeway crossing decisions	Females (%) (n = 37)	Males (%) (n = 63)	All respondents (%)
Safety from crime	64.9	52.4	57.0
Shortest walking distance	13.5	33.3	26.0
Where most people cross	10.8	12.7	12.0
Traffic volumes	8.1	1.6	4.0
Safety from vehicle collision	2.7	0.0	1.0
Total	100	100	100

Table 3.7 Intercept survey respondents' stated frequency of grade-separated crossing facility use, by number of years lived in a city (n = 100; Mngomezulu, 2007)

Number of years lived in a city	*Frequency of grade-separated crossing facility use (%)*						
	Once a day	*Most days of the week*	*Once or twice a week*	*Once or twice a month*	*Less often*	*Never*	*Total*
0–5 years (n=24)	0.0	8.3	12.5	20.8	4.2	54.2	100
6–10 years (n=32)	0.0	6.3	28.1	28.1	18.8	18.8	100
11–20 years (n=33)	6.1	3.0	24.2	18.2	27.3	21.2	100
> 20 years (n=11)	9.1	0.0	36.4	18.2	27.3	9.1	100
All respondents	3.0	5.0	24.0	22.0	19.0	27.0	100

Study 4: observed and surveyed freeway crossing behaviour in Cape Town

Of the crossing pedestrians observed on the selected sections of the N2 and R300 freeways in the second freeway study in Cape Town, 38% crossed at the crossing facilities (with the caveat that the length of observed road sections were not exactly equal), thus exhibiting greater use of crossing facilities than that observed in the earlier arterial study (38% vs. 5–15%). The remaining 62% of crossings were at other points, as illustrated in Figure 3.6. As in the Cape Town arterial study, patterns of crossing away from the facilities are likely to be better explained by the trajectory of pedestrian desire lines, rather than by detour. It was found that despite the greater speed differential and risk associated with freeways, many pedestrians continued to be reluctant to detour even short distances to use a grade-separated facility.

Analysis of the intercept survey data revealed a fairly even spread of respondents across age and gender categories (see Table 3.8). In contrast to the earlier survey, there were a slightly greater proportion of females in the sample (57% males vs. 43% females).

When questioned on why they were crossing at that particular at-grade point, three single or combined most important reasons were given. Table 3.9 presents these findings, indicating that, as in the case of the earlier survey, the most common reason was the desire to walk the shortest route (66%). This was followed by an equal concern for route distance and safety from

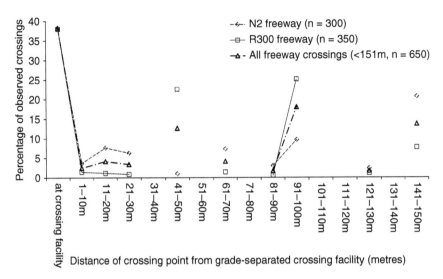

Figure 3.6 Pedestrian crossing distance from grade-separated facility, by freeway study (n = 650; Gabuza, 2008).

crime (21%), and then an equal concern for route distance and crossing in the same place as others (14%). Male respondents reported route distance as the main reason more often than females, and appeared less concerned about crossing where others crossed (suggesting less of a concern for security than amongst females).

When questioned on the most important cause of pedestrian crashes on freeways, 55% of respondents indicated this was drunken pedestrians, followed by drunken motorists (31%), and then poor visibility (15%) (see Table 3.10). Response patterns across the genders were broadly similar, but noticeably different between the at-grade and grade-separated crosser groups. The grade-separated crossers reported greater attribution of collision causes to pedestrian behaviour than in the case of the at-grade crosser group.

Perhaps the most interesting insight derived from the survey concerned the relationship between crossing behaviour and experience of living in a city. Table 3.11 presents findings with respect to the relationship between grade-separated crossing facility use, and the number of years the respondent had lived in a city[13] and gender. Noticeable differences in behaviour patterns can be observed across gender categories (33% of males used grade-separated crossings compared to 45% of females), and significant differences can be observed across categories defined on the basis of experience of city living (90% of respondents who had lived in a city for less than three years crossed at-grade, while 31% of respondents who had lived in a city for more than 15 years crossed at-grade). Of the 401 intercepted respondents who crossed at-grade, 51% had lived in a city for two years or less, whereas amongst respondents who used the grade-separated crossing facility, only 9% had lived in a city

Table 3.8 At-grade and grade-separated crossers, by age and gender (n = 649; Gabuza, 2008)

Age group	At-grade crossers (n = 401)						Grade-separated crossers (n = 248)						Total	
	Females	%	Males	%	Sub-total	%	Females	%	Males	%	Sub-total	%		%
<19 yrs (n = 136)	41	6.3	48	7.4	89	13.7	16	2.5	31	4.8	47	7.2	136	21.0
19–25 yrs (n = 94)	22	3.4	32	4.9	54	8.3	17	2.6	23	3.5	40	6.2	94	14.5
26–34 yrs (n = 117)	40	6.2	43	6.6	83	12.8	17	2.6	17	2.6	34	5.2	117	18.0
35–45 yrs (n = 199)	35	5.4	89	13.7	124	19.1	49	7.6	26	4.0	75	11.6	199	30.7
>45 yrs (n = 103)	16	2.5	35	5.4	51	7.9	26	4.0	26	4.0	52	8.0	103	15.9
Total	154	23.7	247	38.1	401	61.8	125	19.3	123	19.0	248	38.2	649	100

Table 3.9 Intercept survey respondents' stated reason for crossing at-grade, by gender (n = 401; Gabuza, 2008)

Reason for crossing at chosen crossing point	Females (n = 154)		Males (n = 247)		All at-grade crossers	
		%		%		%
Shortest route	92	59.6	172	69.6	264	65.8
Shortest route and safety from crime	29	19.1	53	21.5	83	20.6
Shortest route and crossing with others	33	21.3	22	8.9	55	13.6
Total	154	100	247	100	401	100

for two years or less. The data in the table suggests that time in the city has a relatively smaller impact on the adaption of men's crossing behaviour than women's (42% and 48% of at-grade crossers who had lived in a city for less than three years were female and male respectively, while these proportions were 5% and 26% for at-grade crossers who had lived in a city for greater than 15 years).

Differentiating the effect of time in a city and age on the use of crossing facilities, and their relative strength, is difficult in a small sample survey of this nature. Table 3.12 presents a cross-tabulation of these two variables, and it is apparent that both age and time in a city are inversely related to the likelihood of at-grade crossing. It was noted above that 90% of the 228 respondents who had lived in a city for less than three years crossed at-grade. Of the 136 respondents younger than 19 years, 65% crossed at-grade, while of the 103 respondents older than 45 years, 50% crossed at-grade. Clearly then experience of traffic safety risks, whether it is gained from living in a city or through maturing with age, is of importance in adapting non-compliant crossing behaviour.

Figure 3.7a-b illustrates relationships between crossing behaviour and the demographic variations presented in Tables 3.11 and 3.12. The plot of grade-separated crossers aged less than 19 years, in Figure 3.7b, clearly illustrates that confounding effect of age and time in a city, with respect to understanding impact on spatial non-compliance.

Conclusion: synthesis and implications

Reflecting upon the research hypothesis set out at the start of the section 'Arterial crossing behaviour studies', a synthesis of research findings across the studies in Cape Town (presented in Figure 3.8) indicates that the distribution of points of pedestrian arterial and freeway crossing, in relation to provided

Table 3.10 At-grade and grade-separated crossers' perceived causes of pedestrian–vehicle collisions, by gender (n = 650; Gabuza, 2008).

Perceived cause of pedestrian–vehicle collisions	At-grade crossers (n = 402)						Grade-separated crossers (n = 248)						Total	
	females (n = 154)	%	males (n = 248)	%	Sub-total	%	females (n = 125)	%	males (n = 123)	%	Sub-total	%		%
Drunken peds	62	40.3	108	43.5	170	42.3	90	72.0	94	76.4	184	74.2	354	54.5
Drunken motorists	72	46.8	85	34.3	157	39.1	25	20.0	16	13.0	41	16.5	198	30.5
Poor visibility	20	13.0	55	22.2	75	18.7	10	8.0	13	10.6	23	9.3	98	15.1
Total	154	100	248	100	402	100	125	100	123	100	248	100	650	100

Table 3.11 Grade-separated crossing facility use, by number of years lived in a city and gender (n = 649; Gabuza, 2008).

Gender	Number of years lived in a city								Sub-total		Total
	1–2 yrs		3–6 yrs		7–15 yrs		>15 yrs				
	AGC	GSC	AGC	GSC	AGC	GSC	AGC	GSC	AGC	GSC	
Females	96	8	28	18	20	29	10	70	154	125	279
%	42.1	3.5	27.7	17.8	15.9	23.0	5.2	36.1	55.2	44.8	100
Males	110	14	34	21	52	25	51	63	247	123	370
%	48.2	6.1	33.7	20.8	41.3	19.8	26.3	32.5	66.8	33.2	100
Sub-total	206	22	62	39	72	54	61	133	401	248	649
%	90.4	9.6	61.4	38.6	57.1	42.9	31.4	68.6	61.8	38.2	100
Total	228		101		126		194		649		
%	100		100		100		100		100		

Note: AGC = at-grade crossers; GSC = grade-separated crossers

Table 3.12 Grade-separated crossing facility use, by number of years lived in a city and age group (n = 649; Gabuza, 2008).

| Age group | Number of years lived in a city | | | | | | | | Sub-total | | Total |
| | 1–2 yrs | | 3–6 yrs | | 7–15 yrs | | >15 yrs | | | | |
	AGC	GSC	AGC	GSC	AGC	GSC	AGC	GSC	AGC	GSC	
<19 yrs	59	14	18	14	4	8	8	11	89	47	136
%	25.9	6.1	17.8	13.9	3.2	6.3	4.1	5.7	65.4	34.6	100
19–25 yrs	32	1	7	4	11	8	4	27	54	40	94
%	14.0	0.4	6.9	4.0	8.7	6.3	2.1	13.9	57.4	42.6	100
26–34 yrs	52	1	8	5	12	5	11	23	83	34	117
%	22.8	0.4	7.9	5.0	9.5	4.0	5.7	11.9	70.9	29.1	100
35–45 yrs	41	6	22	10	36	20	25	39	124	75	199
%	18.0	2.6	21.8	9.9	28.6	15.9	12.9	20.1	62.3	37.7	100
>45 yrs	22	0	7	6	9	13	13	33	51	52	103
%	9.6	0.0	6.9	5.9	7.1	10.3	6.7	17.0	49.5	50.5	100
Sub-total	206	22	62	39	72	54	61	133	401	248	649
%	90.4	9.6	61.4	38.6	57.1	42.9	31.4	68.6	61.8	38.2	100
Total	228		101		126		194		649		
%	100		100		100		100		100		

Note: AGC = at-grade crossers; GSC = grade-separated crossers.

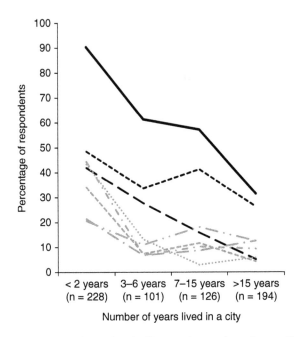

Figure 3.7a Grade-separated crossing facility use, by number of years lived in a city, gender and age group (n = 649; Gabuza, 2008): at-grade crossers

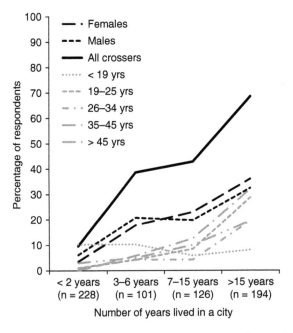

Figure 3.7b Grade-separated crossing facility use, by number of years lived in a city, gender and age group (n = 649; Gabuza, 2008): grade-separated crossers.

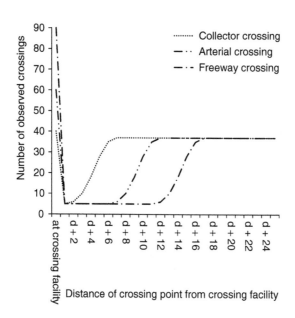

Figure 3.8a Comparison of hypothesized and observed spatial compliance in pedestrian crossing behaviour, by road type: research hypothesis (n = 2,856)

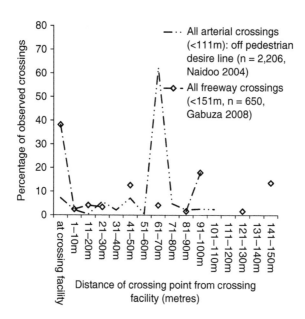

Figure 3.8b Comparison of hypothesized and observed spatial compliance in pedestrian crossing behaviour, by road type: observed pedestrian crossing distances from facility, by road type (n = 2,856)

crossing facilities, does not follow a sigmoidal curve. Significant numbers of pedestrians were observed to cross arterials and freeways unassisted at small distances from crossing facilities. While greater use of freeway crossing facilities was observed relative to arterial crossing facilities, it is posited that patterns of crossing would be better explained by the location of the crossing facilities in relation to dominant pedestrian desire lines. The relatively greater freeway crossing facility use is likely to be associated with, amongst other factors (e.g. the illegality of, and the greater physical difficulty presented by, crossing freeways at-grade), a greater perceived risk associated with the greater speed differential on freeways.

The patterns of pedestrian crossing behaviour reported in this chapter are expected to be context specific. Cities where levels of traffic law enforcement and associated compliance levels are higher than in Cape Town are unlikely to reveal similar unassisted or illegal crossing behaviour. King et al. (2009), for instance, in their study of signalized crossing facilities in Brisbane, reported that illegal crossings accounted for 20% of observed crossings, and Mullen et al. (1990) reported 25% in a study in the US. In comparison, the study findings in Cape Town presented in Figure 3.8 suggest that as much as 62% of crossing on freeways, and 93% of crossing on arterials (off desire lines), may be unassisted or illegal. Even within Cape Town, given findings with respect to the relationship between experience of city living and at-grade freeway crossing, behavioural patterns are likely to vary across different parts of the city on the basis of the different socio-demographic groups they accommodate. Other factors are also likely to play a role in attitudinal and behavioural variation, as found, for instance, by Cho et al. (2009) with respect to the influence of neighbourhood design on perceptions of safety, and by Rosenbloom et al. (2004) with respect to religiosity.

What then are the main implications of the findings presented in this chapter for the formulation of strategies to improve crossing safety by reducing unassisted and illegal pedestrian crossing behaviour?[14] A first, obvious, implication is that the provision of regularly spaced crossing facilities, on their own, is unlikely to lead to significant changes in pedestrian crossing behaviour. The studies in Cape Town suggest that crossing facilities are more likely to be used if they are located on the pedestrian's desire line.[15] Consequently, understanding or estimating pedestrian desire lines and walking trip assignment is more important than understanding detour refusal distances in locating crossing facilities and in attempting to minimize unassisted or illegal crossing patterns. This necessitates that walking be routinely included in travel behaviour analysis, when in the past it has been omitted, and treated as a travel mode like any other. Trip stage and main mode walking trips need to be analyzed in travel surveys, and methods for analysing and predicting walking trip generation, distribution and route choice need to be developed.

A second linked implication – given the finding in Nairobi that pedestrian crossing (refuge) islands and unsignalized zebra crossings on Jogoo Road are ineffective as a means of simultaneously facilitating safe pedestrian crossing and achieving a significant 'no-discomfort' traffic calming effect on speeding motorists – is that consideration should be given to the installation of optimally spaced (i.e. logically linked to walking routes) and appropriately designed raised zebra crossings.[16] In contexts where driver compliance with traffic rules is low and law enforcement is limited, and the urban road speed limit is therefore seldom observed by drivers and the crash risk to non-drivers is high, a possible way forward is to penalize drivers decisively for non-compliance by providing raised crossings as a speed calming and pedestrian safety improvement measure. Raised zebra crossings can also enhance the fluidity of traffic flow by reducing large speed differences between vehicles, thereby leading to improved road capacity. This is one way of improving the safety of pedestrians while at the same time achieving a level of operational efficiency in the use of limited road space. The efficacy of raised zebra crossings *vis-à-vis* other traffic calming measures needs to be tested in Nairobi, however, in order to assess their cost effectiveness and their effect on vehicle-to-vehicle collisions in traffic streams. Other contexts, with greater driver compliance and enforcement capacity, may require less physically self-enforcing measures.

A third implication, given the observed relationship between city living and crossing behaviour in Cape Town, is that key to the formulation of any strategy to improve pedestrian safety will be an acceleration of the learning experience and appreciation of traffic risk. This indicates that education and awareness programmes will be important, in parallel with improved enforcement of traffic rules pertaining to pedestrian crossing. For education and awareness campaigns to be most effective, research into variations in attitudes and behaviour across population segments will be necessary to identify the most appropriate communication medium and targeting of campaigns.

Notes

1 The Nairobi City Council has over the years had a skewed focus on transport infrastructure investments that has tended to promote the needs of motorized travel mode over those of non-motorized road users.
2 The term 'detour refusal rate' is defined in this chapter as the proportion of pedestrians crossing a road at a particular point who are unwilling to extend their walking trip distance in order to utilize a pedestrian crossing facility.
3 For simplification, the figure assumes that the demand for pedestrian crossing is homogeneous along the length of all three classes of road. It should be noted that this demand will of course vary both along the same class of road depending on abutting land use patterns and the socio-economic characteristics of residents, as well as across different classes of road depending on the travel behaviour patterns and law enforcement characteristics of different city contexts.
4 The term 'pedestrian desire line' is defined in this chapter as the route that pedestrians would prefer to take to get from one location to another in order to minimize their travelled path, irrespective of crossing regulations.
5 These did not necessarily fall on the designated crossing points marked on the roadway.
6 Research into road crossing speeds elsewhere has indicated an average value in the range 1.2 metres/second to 1.35 metres/second at busy crossings with a mix of

pedestrian age groups. However, if crossings are less busy, then the mean walking speeds approximating to the free-flow walking speeds on pedestrian courses of 1.6 metres/second can be expected. For disabled people, a more appropriate value is 0.5 metres/second if the needs of most disabled people are to be satisfied (Leake, 1997).

7 At the Muthurwa crossing there is a painted zebra crossing unlike the other survey points, but this is largely disregarded by motorists.

8 The speed limit on urban arterials in Nairobi, and any other cities in Kenya, is set at 50 kilometres/hour.

9 Tumlin (2012), based on road crossing behaviour in North America, observes that driver predictability has a significant influence on the judgement of other road users in terms of when and where to undertake road crossing manoeuvres.

10 It should be noted here, and in other similar charts presented later in this paper, that the percentage value of crossings at different points from the crossing facility is determined either by the length of road section analyzed or the distance between crossing facilities. Percentage values are used in charts to facilitate visual comparisons between roads where crossing volumes are different. The percentage values provided in charts are, therefore, not of great importance from a relative value perspective.

11 Similar observations were made by Masaoe (2011) in the context of Kinondoni Municipality in Dar es Salaam.

12 A larger crossing count study on the R300 freeway, involving 23 count locations over two 1-hour periods in November 2007, corroborates that large numbers of pedestrians cross the freeway (9,032/12-hour period), and provides insight into the timing of crossings (finding in the region of 1,600 crossings in the 06h00–07h00 peak hour, and 450 crossing/hour in the off peak) (Erasmus, 2008). Of the 9,032 crossings per 12-hour observation period, approximately one-third were illegal (Randall Cable, South African National Road Agency, pers comm, 2009). A complementary (n = 117) pedestrian intercept survey provided insight into trip purposes (with social visits, shopping and works trips dominant). A limitation of these data, however, is that they were collected in school vacation time, and therefore omitted school trips.

13 A large proportion of low-income households living in Cape Town are first-generation residents who have migrated from rural districts and small towns in the Eastern Cape (De Swardt et al., 2005).

14 This objective assumes that it is safer for crossing pedestrians to be spatially compliant than spatially non-compliant. A study by Zegeer et al. (2005) found, in the context of the US, that on multi-lane roads with traffic volumes above ±12,000 vehicles/day, the presence of a marked crosswalk alone (without other substantial improvements) was associated with a higher pedestrian crash rate (after controlling for other site factors) compared to an unmarked crosswalk.

15 A 'before-and-after' assessment of the spatial compliance impacts of a pedestrian footbridge constructed on the pedestrian desire line observed on Buitengracht Street westbound (see Figure 3.4a) found virtually no instances of non-compliance in the 'after' data. Data were collected in 2008 and 2011 by Slingers (2012).

16 A study by Mburu (2002) on smaller collector roads in the vicinity of Jogoo Road demonstrated the effectiveness of traffic calming and pedestrian crossing facilities in this context with lessons that can be replicated elsewhere in a developing city with comparable driver behaviour. In this study, it was shown that by constructing short raised road sections, crash risks significantly reduce as driver speeds tend to gravitate around the prescribed speed limits or even lower. Such measures remain very cost effective and could be scaled up on a system-wide basis.

References

Behrens, R. 2002. *Matching networks to needs: travel needs and the configuration and management of local movement networks in South African cities.* Unpublished PhD thesis, University of Cape Town.

Behrens, R. 2005. Accommodating walking as a travel mode in South African cities: towards improved neighbourhood movement network design practices. *Planning Practice and Research,* 20(2): 163–182.

Cho, G., Rodríguez, D. and Khattak, A. 2009. The role of the built environment in explaining relationships between perceived and actual pedestrian and bicyclist safety. *Accident Analysis and Prevention,* 41: 692–702.

Chu, X., Guttenplan, M., and Baltes, M. 2004. Why people cross where they do: the role of street environment. *Transportation Research Record: Journal of the Transportation Research Board,* 1878: 3–10.

CoCT. 2004. *Traffic accidents statistics 2004.* Cape Town: City of Cape Town.

De Langen, M. and Tembele, R. 2001. *Productive and liveable cities: Guidelines for pedestrian and bicycle traffic in African cities.* Lisse: IHE-Delft, A.A. Balkema Publishers.

De Swardt, C., Puoane, T., Chopra, M. and Du Toit, A. 2005. Urban poverty in Cape Town. *Environment and Urbanization,* 17(2): 101–111.

Erasmus, P. 2008. *R300 upgrade pedestrian study.* Report for the South African National Roads Agency prepared by SNA-SSI Joint Venture, Cape Town.

Gabuza, O. 2008. *An analysis of pedestrian crossing behaviour on arterials and freeways.* Master of Engineering (Transport Studies) unpublished 60-credit minor dissertation, University of Cape Town.

Ishaque, M. and Noland, R. 2008. Behavioural issues in pedestrian speed choice and street crossing behaviour: A review. *Transport Reviews,* 28(1): 61–85.

King, M., Soole, D. and Ghafourian, A. 2009. Illegal pedestrian crossing at signalised intersections: incidence and relative risk. *Accident Analysis and Prevention,* 41: 485–490.

Leake, G. 1997. Planning for pedestrians, cyclists and disabled people. In O'Flaherty, C. (ed.), *Transport planning and traffic engineering.* Oxford: Elsevier Butterworth-Heinemann.

Masaoe, E. 2011. *Safety of vulnerable road users in urban centres: The case of Kinondoni municipality.* 5th Africa Transportation Technology Transfer (T2) Conference, Arusha.

Mburu, S. 2002. *Evaluation of traffic calming and pedestrian crossing facilities in Nairobi, Kenya.* MSc Thesis, IHE, Delft, the Netherlands.

Mengot, A. 2012. *The first annual review report of the Africa Road Safety Corridor Initiative: Northern corridor (Kenya – Uganda segment).* Decade of Action for Road Safety 2011–2020, Global Road Safety Facility and the World Bank.

Mngomezulu, T. 2007. *Pedestrian crossing behaviour on freeways.* Unpublished final year Civil Engineering undergraduate thesis, University of Cape Town.

Mullen, B., Cooper, C. and Driskell, J. 1990. Jaywalking as a function of model behaviour. *Personality and Social Psychology Bulletin,* 16(2): 320–330.

Naidoo, K. 2004. *An investigation of pedestrian crossing behaviour and the use of pedestrian crossing facilities.* Unpublished final year Civil Engineering undergraduate thesis, University of Cape Town.

Papadimitriou, E., Yannis, G. and Golias, J. 2009. A critical assessment of pedestrian behaviour models. *Transportation Research Part F,* 12: 242–255.

Ribbens, H. 1990. *Die evaluering van stadsbeplannings- en ontwerpmaatreëls met betrekking tot voetgangerveiligheid in Suid-Afrika.* Unpublished PhD thesis (Town and Regional Planning), Faculty of Engineering, University of Pretoria, Pretoria, South Africa.

Rosén, E., Stigson, H. and Sander, U. 2011. Literature review of pedestrian fatality risk as a function of car impact speed. *Accident Analysis and Prevention*, 43: 25–33.

Rosenbloom, T., Nemrodov, D. and Barkan, H. 2004. For heaven's sake follow the rules: pedestrians' behaviour in an ultra-orthodox and a non-orthodox city. *Transportation Research Part F*, Vol. 7: 395–404.

Sisiopiku, V. and Akin, D. 2003. Pedestrian behaviors at and perceptions towards various pedestrian facilities: an examination based on observation and survey data. *Transportation Research Part F*, 6: 249–274.

Slingers, N. 2012. *The efficacy of pedestrian safety improvements at the intersection of Buitengracht and Coen Steytler Avenue in Cape Town.* Master of Engineering (Transport Studies) unpublished 60-credit minor dissertation, University of Cape Town.

Tumlin, J. 2012. *Sustainable transportation planning: tools for creating vibrant, healthy, and resilient communities.* Hoboken, NJ: John Wiley and Sons.

Zegeer, C., Stewart, J., Huang, H., Lagerwey, P., Feaganes, J. and Campbell, B. 2005. *Safety effects of marked versus unmarked crosswalks at uncontrolled locations: final report and recommended guidelines.* FHWA publication number: HRT-04-100. McLean, VA: Federal Highway Administration.

4 Road safety and non-motorized transport in African cities

Marianne Vanderschuren and Mark Zuidgeest

Road safety in African cities

As early as 1973, road fatalities were labelled an epidemic (British Medical Journal, 1973). Road fatalities and injuries have claimed a total of over 30 million lives worldwide (NGO Brussels Declaration, 2009) with at least 150 road fatalities (all modes included) being registered every hour of every day, totalling 1.24 million people globally per annum. For people aged 15–29, road accidents pose an even greater risk than HIV, AIDS, tuberculosis and malaria (WHO, 2013). In Africa, the majority of people killed on its roads are young breadwinners – 62% are aged between 15 and 44 years, and 3 out of every 4 deaths are males (Jobanputra, 2013; Ogendi et al., 2013; Kopits and Cropper, 2005).

In 2013, the WHO studied overall road safety in several countries in the African region. These countries showed an average road fatality risk that is well above the global average, i.e. 24.1 casualties per 100,000 population in Africa, compared to 17.0 casualties per 100,000 population globally. Figure 4.1 provides an overview of the road fatality rates per 100,000 population for the various continents, and a number of African countries that are included in the city comparison that follows.

North America and Europe score well below the global average road fatality rates, while Africa has the worst overall road fatality rate when comparing continents. African countries included in the comparison all have an overall road fatality risk that is substantially higher than the global road fatality risk. Amongst these, Nigeria and South Africa have the highest road traffic fatality rates (33.7 and 31.9 casualties per 100,000 population, respectively).

For every crash, there are numerous factors contributing to its severity. Speed and aggression, often influenced by gender, age and alcohol abuse, contribute to the negative overall road fatality statistics in Africa (Matzopoulos et al., 2013; Abegaz, 2014; Monteiro et al., 2015). The speed at which a car is travelling influences both crash risk and crash consequences. The effect on crash risk comes mainly via the relationship between speed and stopping distance. The higher the speed of a vehicle, the shorter the time a driver has to stop and avoid a crash, including hitting a pedestrian (McLean et al., 1994).

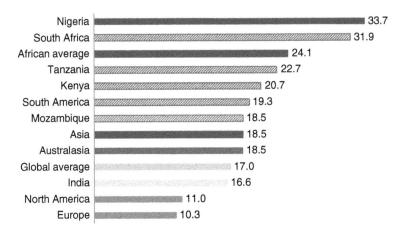

Figure 4.1 Road fatality rates for continents and selected countries (all
modes per 100,000 pop., 2010).

Source: adapted from WHO 2013.[1]

The crossing of mobility corridors (also known as jaywalking) has also
been identified as a major contributing factor. The main contributing factors
to fatalities in South Africa, as surveyed in December 2002, were 47% jaywalking and 30% speeding (Jungu-Omara and Vanderschuren, 2006). A case
study in Cape Town revealed that pedestrians that have lived in the city for
two years or less cross highways significantly more at grade, than using available elevated facilities, i.e. bridges (Behrens, 2010).

Furthermore, pedestrians crossing highways in Cape Town appear to make
a trade-off between the risk of violence and aggression when crossing using
a footbridge and the risk of being hit by a vehicle when crossing at grade
(Sinclair and Zuidgeest, 2016).

It is not only crossing that poses a risk as Ogendi et al. (2013) reveals,
indicating that, although most of the pedestrians were hit while crossing the
road in Nairobi, a sizeable proportion are hit while walking along the road,
or standing by the road.

The underlying causes of crashes within the traffic system are often suggested to emanate from exogenous, rather than endogenous, sources. Many
authors indicate that road geometrics and roadside features are significant
in explaining road crashes (see, for example, Flahaut, 2004; Martin, 2002;
Shankar et al., 1995) in general. However, this is likely to have a direct impact
on NMT as well.

In addition, a city like Nairobi, for example, has a lack of NMT infrastructure provision and motorized transport encroaches into the space that,
ideally, should be available for NMT. Land-use planning and development
control are also weak (Nairobi City County Government, 2015), which results
in a dangerous mix of NMT, public transport and motorized transport.

The fact that mechanical defects, possibly due to lack of inspection and maintenance, in vehicles also plays a role in crash causation is undeniable and is evident from many reports published on road safety (see, for example, ETSC, 2001).

In many cases, safety improvements through vehicle design improvements are easy. For example, redesign of bumpers and the car hood can have a significant reduction on the impact of car–pedestrian crashes, without jeopardizing the structural integrity of the car. In countries with relatively old vehicles (hence old vehicle design), or a large percentage of heavy vehicles, it is expected to see a sizable difference.

Most studies analyzing overall road fatality statistics and the causes of road crashes report on a country level rather than a city level. Moreover, the knowledge on NMT-related fatalities is also limited. The aim of this chapter is to provide more insight into road fatality risks for Africa, and African cities in particular, with a focus on NMT.

Methods

Data collection for this research is based on literature and various secondary data sources. In many cases, municipal databases were accessed. Specific focus was on three case cities: Nairobi, Dar es Salaam and Cape Town. In these three cities, mortuary data, police reports and interviews enriched the findings.

In 2013, the WHO harmonized overall traffic safety risk data provided by different countries, with the aim to standardize a road traffic fatality as 'any person killed immediately or dying within 30 days as a result of a road traffic accident' (Economic Commission for Europe Inter-Secretariat Working Group on Transport Statistics, 2003). The choice of 30 days is based on research that shows that most people who die as a result of a crash succumb to their injuries within 30 days of sustaining them, and that while extension of this 30-day period results in a marginal increase in numbers, it requires a disproportionately large increase in surveillance efforts (WHO, 2013). The ACET research team made an effort to harmonize the road fatality data for African cities, in line with WHO recommendations, and compare city and country police reports with mortuary data. The remainder of this chapter will disaggregate the data into sub-populations (by gender, age, vehicle type, day of the week, etc.) and discuss possible ways of mitigating the road safety burden.

Road traffic risk in selected African cities

When looking, specifically, at overall road fatalities in various global cities, the selected African cities (Maputo, Bloemfontein, Dar es Salaam, Durban, Johannesburg, Cape Town, Port Elizabeth, Nairobi and Lagos) for which data are available score typically higher in terms of average overall road fatalities per 100,000 population than the global average (see Figure 4.2). In fact, most of the selected African cities score higher than the African

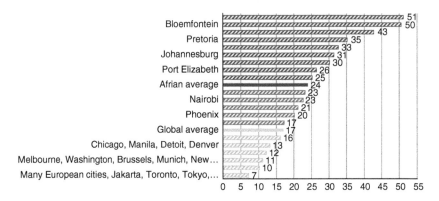

Figure 4.2 Road fatalities per 100,000 inhabitants in various cities (all modes)[2].

average per 100,000 population and are, thus, amongst the most problematic cities in terms of road traffic safety on the continent.

Other than Nairobi, all African cities, for which road fatality data is available, have an overall fatality rate well above the African average. Maputo and Bloemfontein top the list with 51 and 50 fatalities per 100,000 population, respectively. Maputo's fatality rate is also well above Mozambique's average of 18.5 fatalities per 100,000 population, indicating that overall fatality rates in urban areas are higher than in rural areas.

The large South African cities (Pretoria, Durban, Johannesburg and Cape Town) show a similar trend. Fatality rates are similar to the country's average of 31.9 fatalities per 100,000 population. For South Africa, Bloemfontein is an outlier (smaller in size but very high overall fatalities per 100,000 population). The built environment in Bloemfontein has been identified as one of the reasons for the high road fatality rate, as various freeways cross residential areas and most long-distance drivers will reach the city during dusk. Although Nairobi's overall fatality rates per 100,000 population is below the African average, the city still scores above the national (Kenyan) average.

Fatality rates amongst vulnerable road users

Vulnerable road users – pedestrians, cyclists and riders of motorized two- and three-wheelers – constitute more than half (52%) of those killed, with pedestrians alone accounting for 37%. More than 5,000 pedestrians are killed on the world's roads each week (Lee-Brago, 2013). In Lagos, some 50% of all road fatalities are motorcycle drivers and their passengers. Many of these motorcycles are used as an informal public transport and some 70% of drivers indicate that they have been involved in a crash at least once (Oyesiki, 2002).

In low- and middle-income countries, vulnerable road users account for a third of road deaths (WHO, 2009; Peden et al., 2013). This group, especially pedestrians, are over-represented in road traffic fatalities in most Sub-Saharan

Table 4.1 Fatality per mode of transport (%)

Mode	Nairobi	Dar es Salaam	Cape Town
Motor car	30 [26]	20	34
Motorcycle	2 [10]	7	6
Cyclist	3 [5]	6	3
Pedestrian	65 [59]	67	57
Total	100	100	100

Sources: Nairobi police records (2011) [and Nairobi trauma study by Ogendi et al., 2013], Dar es Salaam police records (2007), City of Cape Town Forensic Pathology Laboratory (2011).

African cities. Table 4.1 provides an overview of the fatality rates for different road users in the case cities: Nairobi, Dar es Salaam and Cape Town.

Between 66% and 80% of road fatalities in the three case cities are vulnerable road users. As indicated, the majority of vulnerable road user fatalities are pedestrians. The largest pedestrian fatality burden is suffered in Dar es Salaam with 67%. In Nairobi, between 59% and 65% (depending on the data used) of fatalities are pedestrians, while 57% of fatalities in Cape Town are pedestrians. In cities like Cape Town, pedestrian activity on urban freeways and arterials contribute, significantly, to pedestrian fatality rates. The ten most hazardous locations identified in the city are all urban freeways or arterials.

The results from the three case cities are supported by previous research. For example, pedestrians in Nairobi constituted the largest number of road traffic fatalities (65%), followed by cyclists (3%) and motorcyclists (2%), in a study conducted by Khayesi (1997). A recent study by Ogendi (Ogendi et al., 2013) analyzed road trauma admissions to the largest hospital in Nairobi, Kenyatta National Hospital. A total of 253 road traffic trauma cases were admitted during the study period of 3 months in 2011. Of these admissions, 70% were injured in crashes within Nairobi City. The majority of cases were pedestrians (59%), with motor vehicle passengers the second most common case (24%), followed by motorcyclists (10%), cyclists (5%) and drivers (2%). Cars and buses (matatus) were almost equally represented in the accidents involving motor vehicles with 39.4 versus 35.5%.

Dar es Salaam has the highest concentration of traffic, as well as traffic crashes in Tanzania, especially crashes involving pedestrians, cyclists and motorcyclists. In police data collected in 2007 and 2008 (Nyoni and Masaoe, 2011), it was reported that 67% of fatalities were pedestrians, 6% cyclists and 7% motorcyclists. Drivers and passengers make up just over 20% of fatalities and 48% of non-fatal injuries (Nyoni and Masaoe, 2011).

In Cape Town, NMT users, of which the majority are pedestrians, account for 60% of all fatalities. Although the fatality rate for cyclists seems low (3%), when related to the modal split, this is not the case. Cycling accounts for less than

1% of the trips in Cape Town (NHTS, 2013). On a national level (South Africa), although only 0.8% of trips are by bicycle (NHTS, 2003), the absolute number of fatal bicycle accidents are, for example, twice the number in The Netherlands, a nation where, on average, 27% of trips are bike-based (WHO, 2009).

Gender of pedestrian fatalities in African cities

Studies indicate that males, both children and adults, make up a high proportion of pedestrian deaths and injuries. Although more females walk and use public transport, female pedestrian fatalities represent a minority of crashes in the US and account for 32% of all pedestrian fatalities (NHTSA, 2004). In the three African cities studied, only 20% of pedestrian fatalities are female in Cape Town and Dar es Salaam, while 25% of Nairobi's pedestrian fatalities are female.

Age and road traffic fatalities

In developed countries, older pedestrians are more at risk, while in low-income and middle-income countries, children and young adults are often affected (www.who.int/mediacentre/factsheets/fs358/en/). The same holds for Nairobi, where 47% of pedestrian fatalities are below the age of 30 years old. However, in Cape Town the age distribution for pedestrian fatalities is very different. Only 28% of pedestrian fatalities in Cape Town are children and young adults below the age of 29 years. An equal number of fatalities is found in the age group between 45 and 59 years old, while the largest number of pedestrian fatalities in Cape Town is found in the age group between 30 and 44 years old (see Figure 4.3). Unfortunately, no age-based pedestrian fatality information was available for Dar es Salaam.

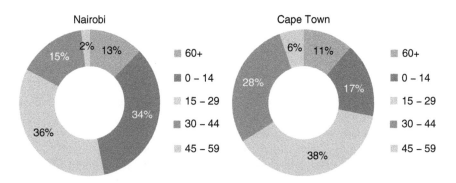

Figure 4.3 Age split for pedestrian fatalities in various cities.

Sources: Nairobi (Ogendi, see Chapter 5), Cape Town (City of Cape Town Forensic Pathology Laboratory, 2014).

Road fatalities on different days of the week

Analysis of overall fatalities during different days of the week (see Table 4.2) indicates that the road safety burden is highest over the weekend (starting Friday afternoon). The case study cities confirm this, i.e. fatalities on Saturday are highest in Nairobi, with 25% of all fatalities, followed by Cape Town with 20% of all fatalities on that day. Saturdays and Sundays combined had a total of 42% of all fatalities in Nairobi and 38% in Cape Town. The skewing towards the weekend is likely caused by alcohol abuse. In a study by the Medical Research Council (2006), pedestrians had the highest percentage of cases that tested positive for alcohol (60%) at the time of the collision, as well as the highest levels of alcohol consumption (on average four times the legal limit). It appears that in Dar es Salaam particularly, the burden on the weekend is not significantly higher than on other days of the week. Reasons for this have not been identified, but could be related to different alcohol consumption patterns and/or motorization levels.

Vehicles causing (pedestrian) fatalities

One of the main factors contributing to the increase in global road crash injuries is the growing number of motor vehicles. The growth in vehicle numbers and size increases the exposure to the risk. Aspects of vehicles that increase the likelihood of accidents when travelling are so-called vehicle factors, which include unroadworthy vehicles, worn-out tyres, wrong air pressure, overloading of vehicles and faulty brakes and vehicle lights (Jungu-Omara and Vanderschuren, 2006). A study by Moodley and Allopi (2008) of vehicular factors leading to crashes in South Africa indicated that the contribution of vehicle 'defects' to fatal road crashes varied between 5% and 17%.

Table 4.2 Fatality per day of the week in various cities (all modes in %)

Day of the week	Nairobi	Dar es Salaam	Cape Town
Monday	11	14	13
Tuesday	11	15	12
Wednesday	11	14	12
Thursday	9	13	10
Friday	16	14	15
Saturday	25	15	20
Sunday	17	15	18
Total	100	100	100

Sources: Nairobi (Ogendi et al., 2013), Dar es Salaam (Masaoe, 2010), City of Cape Town (Jobanputra, 2013).

The type of vehicle involved in accidents in Cape Town does not simply reflect the vehicle fleet registered. The likelihood of a particular vehicle type having an accident is also influenced by vehicle kilometres driven (data which is unfortunately not kept in Cape Town), leading to a vehicle type density on the road. Furthermore, driver behaviour (speeding and alcohol abuse) is a contributing factor. Table 4.3 provides an overview of road crashes per vehicle type.

Although most crashes involve sedans/station wagons, the risk factor is generally low. Minibus taxis (MBTs) have the highest road crash risk, more than five times the level for sedans/station wagons. Buses, especially articulated buses, also have a high road crash risk. The vulnerable vehicle group that was identified in the data, i.e. motorbikes, have the same risk factor as sedans/station wagons.

The question remaining is what vehicles cause fatalities amongst other vulnerable road users. An example is provided by the WHO (2004) in their 'World Report on Road Traffic Injury Prevention' where they state that vehicle design can have considerable influence on injuries, and its contribution to crashes through vehicle design is around 3% in high-income countries, about 5% in Kenya and 3% in South Africa (Jobanputra, 2013).

Table 4.3 Types and number of registered vehicles involved in crashes (2001)

Vehicle type	No. of vehicles involved in accidents	%	No. of registered vehicles	Risk per registered vehicle type
Sedan/station wagon	85,825	66.0	551,892	0.16
Light delivery vehicle	18,227	14.0	147,306	0.12
Combi/minibus taxi	8,428	6.5	10,254	0.82
GVM >3500 kg	5,274	4.1	18,090	0.29
Motorbike	3,073	2.4	19,717	0.16
Bus	1,394	1.1	5,202	0.27
Articulated trucks	1,212	0.9	2,601	0.47
Other	1,361	1.0	14,515	0.09
Unknown	5,254	4.0	3,240	1.62
Total	130,048	100.0	772,817	

Source: Adapted by Jungu-Omara and Vanderschuren (2006) from Cape Town Metropolitan Council data (www.capetown.gov.za), 2003.

Pedestrian fatalities are the only vulnerable road user mode with sufficient data. Based on the City of Cape Town Forensic Pathology Laboratory (2014), it appears that in 24% of the cases, sedans are the cause of a pedestrian fatality, while bakkies (light delivery vehicles) are responsible for 7% of pedestrian fatalities. MBTs are only responsible for 5% of the pedestrian fatalities (see Figure 4.4). Per registered vehicle, however, the risk of being hit by a MBT is 11 times higher than the risk of being hit by a sedan vehicle. For almost 60% of pedestrian fatalities, no vehicle indication is provided. Vanderschuren and Jobanputra (2010) found that the South African Police Services (SAPS) do not see the administration related to road crashes as one of their main responsibilities. Official Accident Reports (OARs) have many missing fields and the forwarding of data is often delayed. Part of the unknown data is also hit and run crashes, where the body of a pedestrian is found on the side of the road.

Mitigating the African road safety burden

Given the road traffic safety figures, as well as the key driving factors discussed previously, a strategy to mitigate the overall road safety burden is needed in African cities, in general, and in the three case cities, Nairobi, Dar es Salaam and Cape Town, in particular. Given the large portion of vulnerable road users, particular attention needs to be given to these groups of road users. The responsibility of achieving traffic safety goals is shared between the road users and system owners (i.e. those responsible for the road infrastructure, vehicle manufacturers, designers and authorities with an involvement and interest in safety). Countries such as Sweden have adopted the ambitious long-term target of zero fatalities as a consequence of road traffic incidents. Amongst the strategic principles to achieve this 'Vision Zero' goal, are that the traffic system must be adapted to take better account of the needs, mistakes and vulnerabilities of road users by appropriate design, and that speed is the most

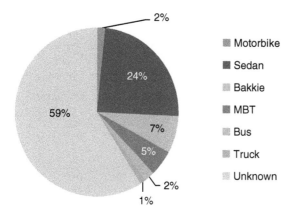

Figure 4.4 Vehicle types causing pedestrian fatalities in Cape Town.

Source: based on City of Cape Town Forensic Pathology Laboratory (2014).

important regulating factor for safe road traffic (SNRA, 2003). The Dutch 'Sustainable Safety' approach is similar, and presents a multi-faceted error management system (Wegman and Oppe, 2010).

The close interaction between the three contributing factors (road, vehicle and human) makes it difficult to isolate a single cause for a crash. Despite this, in-depth crash investigation, or accident reconstruction, attempts to provide this background information, through the use of experts with different disciplinary backgrounds, and in accordance with a theoretical reference frame. It is routinely carried out in most developed nations, but is sadly lacking in Africa, except for the most severe or politicized cases (Jobanputra, 2013).

Typically, crash investigations treat accidents as events. There are three temporally separated phases: 'pre-crash', 'crash' and 'post-crash', based on the systems approach developed by Haddon, as early as 1983. The knowledge gained from studies such as these can be usefully employed for taking safety counter-measures for each of the component factors of the road system and their interaction, which is dynamic and often complex, and the reason why causes (faults) are split into these component parts (see Table 4.4).

Several countries throughout the world have adopted 'The 4 "E"s strategy', with South Africa being among them. The 4 Es are Enforcement, Education, Engineering and Evaluation. Furthermore, the UN launched

Table 4.4 Haddon's road safety matrix

PHASE	GOAL	FACTORS		
		Human	*Vehicle and equipment*	*Road environment*
Pre-crash	Crash prevention	Information Attitudes Impairment Police enforcement	Roadworthiness Lighting Braking Handling Speed management	Road design/ layout Speed limits Pedestrian facilities
Crash	Injury prevention during the crash	Use of restraints Impairment	Occupant restraints Other safety devices Crash protective design	Crash protective roadside objects
Post-crash	Sustaining life	First-aid set Access to medics	Ease of access Fire risk	Rescue facilities Congestion

Source: Haddon (1983).

the Decade of Action for Road Safety 2011–2021 during its General Assembly in March 2010 (UN, 2010), identifying the five pillars of road safety. Although there are variations between these approaches, they are often interlinked or a subset of each other. The 4 Es can be linked to Haddon's crash theory, as well as the five pillars of road safety, and can be used to design mitigation measures and strategies in each crash phase, as illustrated in Table 4.5.

Each one of Haddon's category mitigation measures and categories can be defined using the 4 Es. For example, potential pre-crash safety counter-measures can assist in combating the road safety risk in the African context. The provision of better education and information, as well as increased enforcement, can change people's attitudes. The Western Cape Province in

Table 4.5 Haddon's crash theory, the five pillars and the 4 E approach for road safety[3]

	Pre-crash	*Crash*	*Post-Crash*
Enforcement	• Policy, strategy and legislation P_3 • Policing $P_3 P_4$	• Speed limits P_2 P_4	• Road safety management P_1 • Congestion management P_1
Education	• Safer road users P_4 • Information provisior P_4 • Attitude based campaigns P_4 • Impairment awareness • Handling/driver training P_4	N/A	• First-aid set P_5 • Access to emergency services P_5
Engineering	• Safer roads and mobility P_2 • Safer vehicles P_3 • Speed management/ limits P_1 • Road design P_2 • NMT facilities P_2	• Restraints/safety devises P_3 • Crash protective design (including for pedestrians P_3 • Crash protective roadside objects P_2	• Ease of access to crash P_5 • Fire risk control P_5
Evaluation	• Road safety audits P_2 • Time to collision	• Phone or on-board data analysis	• In-depth crash investigation • In-depth stats analysis • Policy, strategy and legislation P_5

South Africa has been able to reduce fatalities by almost 30% over recent years, through increased and improved enforcement. Part of this was a name-and-shame campaign in the newspapers. Members of Kenya's social media networks have embraced a similar idea. They are tweeting about accidents they witness, advising caution, and outing drivers who drive dangerously.

Road-worthiness of vehicles is another issue in the African context, leading to road safety risks for all, including vulnerable road users. Problems that occur are faulty brakes, lights, tyres etc., not only endangering the driver and passengers, but also other road users, such as crossing pedestrians. Although a roadworthiness check is required regularly, in some cases, for example, for professional vehicles in countries such as Kenya and South Africa, private vehicles are not checked regularly. Moreover, fraud and corruption are jeopardizing the roadworthiness of professional vehicles.

Road design in Africa is based on US design standards and is largely car orientated, with the implication that there is hardly any attention for NMT in road design. More recently, however, the US has developed and improved NMT design standards. Unfortunately, these have not filtered through into the African context, as yet. South Africa, however, has recently updated NMT facility guidelines with the intent to improve NMT infrastructure implementation (Vanderschuren et al., 2014). The development of NMT design standards and facility implementation can reduce the road safety risk in Africa, specifically for NMT.

Secondly, crash-based safety counter-measures for humans, vehicles and the road environment are all focused on the reduction of the crash impact. For NMT, there are high-tech solutions, such as pedestrian airbags on the bonnet (hood) of vehicles. Currently, these types of high-tech measures are not implemented in the African context. In addition, certain types of cars, which have been banned from Europe's streets because of posing a NMT safety hazard, are still sold widely in Africa.

Thirdly, post-crash-based safety counter-measures are focused on the access to emergency facilities. As pedestrians in Africa are mostly captives, i.e. forced to walk, due to a lack of sufficient income, they often also lack financial and practical access to emergency facilities, as public emergency facilities are scarce. In the Western Cape, for example, over 30% of pedestrians die outside the emergency service areas, i.e. areas in which emergency services can get you to emergency care facilities within the so-called 'Golden Hour' (Vanderschuren and McKune, 2015).

Conclusion

Based on the findings in this chapter, it can be concluded that there is a need to reduce the road traffic injury burden in Africa, especially for NMT users. A multitude of factors influence the road safety burden in African countries and cities. Vehicle ownership and speed, as well as substance abuse, have been identified as having a significant influence on the road safety risk. Jaywalking, often caused by security risks and bad land-use planning, is also an issue.

Age, gender and the day of the week have also been identified internationally as playing a major role. The data presented in this chapter confirms that this is also the case in Africa. Furthermore, although NMT does not cause a threat to other road users, they represent the majority of fatalities on the African continent.

The causality of fatalities has been categorized in different contributory factors (road, vehicle and human) by Sabey and Staughton as early as 1975. In Africa, knowledge on the role that the road environment plays in fatality rates is lacking. Moreover, although some vehicle manufacturing happens in Africa, the R&D is done overseas. The African vehicle industry is, therefore, a follower and not an early adopter of road safety technology.

Reducing the road safety risk on the African continent requires holistic mitigation strategies. Following Haddon's (1968) findings, these mitigation strategies can typically involve a 'pre-crash,' 'crash' or 'post-crash' approach. NMT specific strategies typically focus on the provision of information and education with the aim to change road user attitudes; improving enforcement, especially focused on the reduction of identified illegal behaviour (speeding, alcohol abuse and jaywalking); and reducing the encroachment onto NMT facilities by other modes. The provision of improved NMT facilities (see also Chapters 7, 8, 12 and 13) and the adoption of technologies that can reduce the impact of crashes for NMT users are other possible mitigation strategies. Finally, the improvement of post-crash care, through the provision of and improved access to emergency facilities, is a further potential mitigation strategy that could be implemented.

Overall, Africa will only be successful in reducing the road safety burden, especially for NMT users, if multiple mitigation strategies are implemented including improved land-use and transportation planning, the provision of NMT specific facilities, and better road safety education and enforcement. To develop these multiple mitigation strategies disaggregated data needs to be collected and analysed to create the scientific base for appropriate action.

Acknowledgements

The authors would like to express their gratitude to all researchers that have contributed to African Centre of Excellence for Studies in Public and Non-Motorised Transport's (ACET) research in the field of road safety: Dr Rahul Jobanputra, Dr Japheths Ogendi, Dr Wilson Odero, Dr Estomohi Masaoe, Prof Winnie V. Mitullah and Prof Roger Behrens.

Notes

1 Any person killed immediately or dying within 30 days after a road traffic crash. In countries where the year or definition of road fatalities is different, a hominization factor was calculated and applied (see methodology for more information).
2 Various local sources were used to establish city-based fatality rates. Although an attempt was made to use reliable sources, this could not always be verified by the

authors. Furthermore, the data has not been standardized for a particular year. Sources: Newman and Kenworthy (1999); Cape Metropolitan Council (CMC) (2000); Pladsen (2002); South African Cities Network (SACN) (2004); Nigerian Bureau of Statistics (2009); UNRSC meeting, Mozambique in 2013.
3 The five UN pillars of road safety are indicated using P1–P5 (P1 = Road safety management, P2 = Infrastructure, P3 = Safe vehicles, P4 = Road user behavior, and P5 = Post-crash care.

References

Abegaz, T, Berhane, Y., Worku, A., Assrat, A., and Assefa, A. 2014. Effects of excessive speeding and falling asleep while driving on crash injury severity in Ethiopia: a generalized ordered logit model analysis. *Accident Analysis and Prevention*, 71: 15–21. doi: 10.1016/j.aap.2014.05.003.

Behrens, R. 2010. Pedestrian arterial and freeway crossing behaviour in Cape Town: observations and implications. 12th World Conference on Transport Research, Lisbon.

British Medical Journal. 1973. Road accidents epidemic, 17 February 1973, 370–371.

Cape Metropolitan Council (CMC). 2000. *Cape metropolitan area: road traffic accident statistics 1998*. Cape Town: Directorate, Transportation and Traffic, Cape Metropolitan Council, p. 26.

City of Cape Town Forensic Pathology Laboratory. 2011. Database. Cape Town.

City of Cape Town Forensic Pathology Laboratory. 2014. Cape Town Wide Database.

Dar es Salaam police records. 2007. Database. Dar es Salaam.

Economic Commission for Europe Inter-secretariat Working Group on Transport Statistics. 2003. *Glossary of transport statistics, 3rd edition*. New York, NY: United Nations Economic and Social Council (TRANS/WP.6/2003/6).

European Transport Safety Council (ETSC). 2001. *The role of driver fatigue in commercial road crashes*. Brussels.

Flahaut, B. 2014. Impact of infrastructure and local environment on road unsafety. Logistic modeling with spatial autocorrelation. *Accident Analysis and Prevention*, 36(6): 1055–1066.

Haddon, W. 1968. The changing approach to the epidemiology, prevention and amelioration of trauma: the transition to approaches etiologically rather than descriptively based. *American Journal of Public Health*, 58(8), pp. 1431–1438.

Jobanputra, R. 2013. An investigation into the reduction of road safety risk in Cape Town through the use of microscopic simulation modelling. Thesis submitted for the Degree of Doctor of Philosophy in Civil Engineering, University of Cape Town.

Jungu-Omara, I.O. and Vanderschuren, M.J.W.A. 2006. Ways of reducing accidents on South African roads. Proceedings of the 25th Southern African Transport Conference (SATC), Pretoria, ISBN 1-920-01706-2.

Khayesi, M. 1997. Livable streets for pedestrians in Nairobi: the challenge of road traffic accidents. *World Transport Policy and Practice*, 3(1): 4–7.

Kopits, E. and Cropper, M. 2005. Traffic fatalities and economic growth. *Accident Analysis and Prevention*, 37(1): 169–178. Doi:10.1016/j.aap.2004.04.006.

Lee-Brago, P. 2013. 270,000 pedestrians die yearly in accidents – WHO. Press release, Geneva, May 2013.

Martin, J.L. 2002. Relationship between crash rate and hourly traffic flow on interurban motorways, *Accident Analysis and Prevention*, 34(5): 619–629.

Masaoe, E. 2010. Road safety in Dar es Salaam. ACET Working Paper Number WP13.01.

Matzopoulos, R., Lasarow, A. and Bowman, B. 2013. A field test of substance use screening devices as part of routine drunk-driving spot detection operating procedures in South Africa. *Accident Analysis and Prevention*, 59: 118–124.

McLean, A.J., Anderson, R.W., Farmer, M.J., Lee, B.H. and Brooks, C.G. 1994. Vehicle travel speeds and the incidence of fatal pedestrian collisions, Volume 1. Canberra: Federal Office of Road Safety (CR 146).

Medical Research Council (MRC). 2006. *Keeping pedestrians safe: crime, violence and injury lead programme.* Cape Town: Medical Research Council.

Monteiro, N.M., Balogun, S.K., Kote, M. and Tlhabano, K. 2015. Stationary tailgating in Gaborone, Botswana: the influence of gender, time of day, type of vehicle and presence of traffic officer. *IATSS Research*, 38(2): 157–163.

Moodley, S. and Allopi, D. 2008. An analytical study of vehicle defects and their contribution to road accidents. 27th South African Transport Conference (SATC), Pretoria, July 2008, 469–479. ISBN: 978-1-920017-34-7.

Nairobi City County Government. 2015. Non-motorised transport policy – Towards NMT as the mode of choice. Kenya.

Newman, P. and Kenworthy, J. 1999. *Sustainability and cities: overcoming automobile dependence.* Washington, DC: Island Press, pp. 344–345.

NGO Brussels Declaration. 2009. Available at: www.who.int/roadsafety/ministerial_conference/ngo_declaration_full.pdf.

National Household Travel Survey (NHTS). 2003. South African Wide Database. Pretoria, South Africa: Statistics South Africa.

National Household Travel Survey (NHTS). 2013. South African Wide Database. Pretoria, South Africa: Statistics South Africa.

NHTSA. 2004. Traffic safety facts 2002: pedestrians. US Department of Transportation. Available at: www-ndr-nhtsa.dot/pdf/ndr-30/NCSA/TSF2002/2002pedfacts.pdf

Nigerian Bureau of Statistics. 2009. UNRSC meeting, Mozambique, Progress in road safety in the Decade of Action for Road Safety 2011–2020. Geneva.

Nyoni J.E. and Masaoe E.N. 2011. Factors contributing to high frequency of vulnerable road user injury in Dar es Salaam. Proceedings of the 30st South African Transport Conference (SATC), Pretoria, July 2011, pp. 215–223. ISBN: 978-1-920017-51-4.

Ogendi J., Odero, W., Mitullah, W. and Khayesi, M. 2013. Pattern of pedestrian injuries in the city of Nairobi: implications for urban safety planning. *Journal of Urban Health: Bulletin of the New York Academy of Medicine.* Doi:10.1007/s11524-013-9789-8.

Oyesiki, O.K. 2002. Policy framework for urban motorcycle public transport system in Nigerian cities. 10th Codatu Conference, Mobility for All, Lomé, Togo.

Peden M., Kobusingye, O., Monono, M.E. 2013. Africa's roads – the deadliest in the world. Editorial, South African Medical Journal, 103(4): 228–229. Doi:10.7196/SAMJ.6866.

Pladsen, K. 2002. Traffic fatalities increasing in poor countries. Sustainable Mobility News, World Business Council for Sustainable Development.

Sabey, B.E. and Staughton, G.C. (1975). Interacting roles of road environment vehicle and road user in accidents. Zagreb: Hrvatsko društvo za ceste – Via Vita.

Shankar, V., Mannering, F.L. and Barfield, W. 1995. Effect of roadway geometrics and environmental factors on rural freeway accident frequencies. *Accident Analysis and Prevention*, 27(3): 371–389.

Sinclair, M. and Zuidgeest, M. 2016. Investigations into pedestrian crossing choices on Cape Town freeways. *Transportation Research Part F: Traffic Psychology and Behaviour*, volume 42, part 3, October, pp. 479–494.

South African Cities Network (SACN). 2004.. State of the Cities Report 2004. ISBN 620-31150-9; Available at: http://theafricaneconomist.com/.

Swedish National Road Administration (SNRA). 2003. "Vision Zero" – from concept to action. Borlange: Swedish National Road Administration.

Turner, J. and Fletcher, J. 2008. TI-UP enquiry: gender and road safety. Note prepared for TI-UP, May 2008.

United Nations (UN). 2010. UN General Assembly Resolution A/RES/64/255. Improving Global Road Safety. Available at: www.un.org/en/ga/search/view_doc. asp?symbol=A/RES/64/255.

United Nations Road Safety Collaboration (UNRSC). 2013. 17th Meeting of the United Nations Road Safety Collaboration, 14–15 March 2013. Geneva, Switzerland.

Vanderschuren, M. and Jobanputra, R. 2010. Safely Home Project report phase II: baseline study, study for the Provincial Government of the Western Cape. Safely Home Program, November 2010.

Vanderschuren, M. and McKune, D. 2015. Emergency care facility access in rural areas within the golden hour?: Western Cape case study. *International Journal of Health Geographics,* 14(5) (Online: 16 January 2015). DOI: 10.1186/1476-072X-14-5.

Vanderschuren, M., Phayane, S., Taute, A., Ribbens, H., Dingle, N., Pillay, K., Zuidgeest, M., Enicker, S., Baufeldt, J. and Jennings, G. 2014. NMT facility guidelines, 2014 – policy and legislation, planning, design and operations. Pretoria: Department of Transport.

Wegman, F. and Oppe, S. 2010. Benchmarking road safety performances of countries. *Safety Science,* 48, pp. 1203–1211.

World Health Organization (WHO). 2004. World report on road traffic injury prevention. Summary. Geneva: World Health Organization. ISBN 92 4 156260 9.

World Health Organization (WHO). 2009. Global status report on road safety: time for action. Geneva: World Health Organization. ISBN 978 92 4 156384 0.

World Health Organization (WHO). 2013. Global status report on road safety 2013: supporting a decade of action. Geneva: World Health Organization. ISBN 978 92 4 156456 4.

5 Types of injuries and treatment of pedestrians admitted to a referral hospital in Nairobi City, Kenya

Japheths Ogendi

Introduction

The commonly analyzed burden of road traffic injuries is at national or urban level, using road traffic fatalities per 100,000 population or 100,000 vehicles (for instance, see Chapter 4 in this book). While this scale of analysis is relevant for revealing trends and magnitude at a national level, it does not reveal the burden on specific sectors or households. The ideal of preventing road traffic collisions affecting pedestrians from occurring is not always achieved in real life. As a result of a crash, pedestrians suffer different types of injuries, and require post-crash care and rehabilitation. One of the settings that we can use to understand the burden of road traffic collisions is post-crash care. Post-crash response consists of pre-hospital, hospital and rehabilitation components (Mock et al., 1999; Van Rooyen et al., 1999; Mock et al., 2003; Peden et al., 2004; World Health Organization, 2013).

This chapter contributes to an understanding of the burden of pedestrian road traffic injuries on post-crash response or care by focusing on the hospital phase for injured pedestrians. The chapter examines demographic characteristics, types of injuries and length of hospital stay of pedestrians who were admitted to Kenyatta National Hospital, a referral hospital in Nairobi.

Methods

This chapter is based on a PhD study conducted by the author (Ogendi, 2014), which collected data on several variables from different sources, including routinely collected data by the traffic police and interviews. Data on overall trends and situations of pedestrian road safety has been published in another paper (Ogendi et al., 2013). Therefore, this chapter focuses on three variables related to age and sex, types of injuries and length of hospital stay of pedestrians, who were admitted to Kenyatta National Hospital during a three-month period from 1 June 2011 to 31 August 2011.

Kenyatta National Hospital (KNH) is the largest hospital in the city of Nairobi. It covers an area of 45.7 hectares and is a host to other institutions such as (i) the College of Health Sciences (University of Nairobi); (ii) the Kenya Medical Training College; (iii) Kenya Medical Research Institute; and (iv) National Public Health Laboratory Services. Altogether, Kenyatta National Hospital has 50 wards, 20 outpatient clinics, 24 theatres (16 specialized), and an Accident and Emergency Department. The hospital has a total bed capacity of 1,800.

Data on injured patients admitted to Kenyatta National Hospital was gathered over the three-month period. Information on characteristics of pedestrians injured in road traffic crashes was obtained from road trauma casualties admitted to Kenyatta National Hospital. All road traffic injury admissions were identified, every morning, by the nursing officers working in the wards and research assistants by checking case notes and files of admissions to the wards. Before conducting the interview, informed consent was sought from the road traffic crash injury patient or, where this was not possible, because of the severity of the injury or age, from the appropriate caretaker or parents. Only patients who gave consent to gather information, either by themselves or respective caretakers or parents, were enrolled. The information collected included the patient's demographic characteristics; date and day of admission; admission outcome and date; category of road user injured; place and time of the crash; day of the week the crash occurred; duration and outcome of admission; class of vehicle involved; conflict type for pedestrians; and the main type of injury that resulted in admission, as reported in treatment files.

Descriptive statistics were generated for demographic characteristics, duration of admission and nature of injury. Continuous variables were summarized using means and median and standard deviation. Analysis was based on frequency tabulation and group comparisons. The mean length of stay (LOS) was compared between that for pedestrians against all other categories of road users and all other road user categories combined, using analysis of variance (ANOVA), as described by Anderson et al. (1994). A p-value of less than 0.05 was considered significant. The results are presented and discussed in the next section with respect to demographic characteristics, types of injuries and length of hospital stay for pedestrians who were admitted to the hospital.

Demographic characteristics of patients

A total of 176 road traffic trauma cases, injured in crashes which occurred in Nairobi City, were admitted to Kenyatta National Hospital between 1 June and 31 August 2011 (see Figure 5.1). Pedestrians constituted the highest proportion (59.1%) of the road traffic injury admissions, followed by passengers (24.4%), motorcyclists (9.7%), bicyclists (5.1%) and drivers (1.7%) (see Figure 5.2).

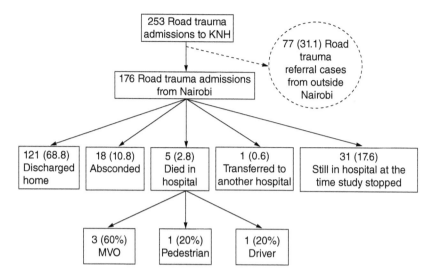

Figure 5.1 Road traffic injury patients admitted at Kenyatta National Hospital, June–August 2011 KNH (n = 103).

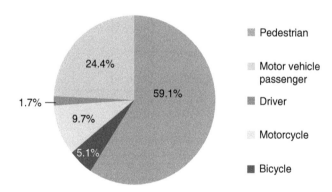

Figure 5.2 Proportion of road traffic injury admissions by road user category, Kenyatta National Hospital, June–August 2011.

Age and sex of admitted pedestrians

Male pedestrians were disproportionately represented, compared to females, in all the age groups with the highest male to female ratio observed in the age bracket of 30–44 (male:female ratio = 8.35:1), followed by those in the age group of 45–59 years (male:female ratio = 4.3:1). A test of significance indicated that the total number of males was only marginally greater ($p = 0.0496$) in all the age groups combined, but significantly greater in the age groups 15–29 ($p = 0.00035$) and 30–34 ($p = 0.0274$), respectively (see Table 5.1).

The analysis also revealed that people aged 15 to 44 years had the heaviest burden of pedestrian injuries: this age range contributed 69.9% of all pedestrians admitted to Kenyatta National Hospital during the study period.

Types of injuries sustained by pedestrians

Information on the type of injury sustained by admitted pedestrians indicated that most of the injuries (65 or 67.7%) occurred to both the upper and lower limbs, and ten (10.4%) to the head and neck regions. Multiple injuries were reported in nine (9.4%) cases (see Table 5.2).

Table 5.1 Distribution of admitted pedestrians by age and sex KNH (n = 103)*

Age group (years)	Amount by gender			Ratio of male to female	P-value
	Total n (%)	Male n (%)	Female n (%)		
0–14	13 (12.6)	10 (76.9)	3 (23.1)	3.3:1	0.2724
15–29	35 (34.0)	21 (60.0)	14 (40.0)	1.5:1	0.0035
30–44	37 (35.9)	33 (89.2)	4 (10.8)	8.35:1	0.0274
45–59	16 (15.5)	13 (81.2)	3 (18.8)	4.3:1	0.6552
60+	2 (1.9)	2 (100.0)		2:1	0.5900
All ages	103 (100)	79 (76.9)	24 (23.1)	3.3:1	0.0496

*Information on age was not available for one pedestrian.

Table 5.2 Admitted pedestrians, Kenyatta National Hospital, June–August 2011

Injury type by body region	Frequency	Percentage
Limb injury (upper and lower limb)	65	67.7
Head and neck	10	10.4
Multiple	9	9.4
Abdomen including lumbar spine and pelvic contents	5	5.2
Face	3	3.1
Thorax including dorsal spine	3	3.1
Unspecified	1	1.1
Total	**96**	**100**

Length of hospital stay

Of the total 4,180 hospital bed days by road traffic injury admissions to KNH during the study period, 2,573 (62%) days were taken up by pedestrians (see Table 5.3). This was 1.6 times higher than that for all other categories of road users combined (2,573 versus 1,607), 2.5 times higher than for motor vehicle passengers (2,573 versus 1,025), which were next when road traffic injury admissions were ranked in order of the number of bed days spent in hospital by each category of road users, and 39 times higher than the total number of bed days by drivers, who had the least total number of hospital bed days (2,573 versus 66). The LOS for admitted pedestrians ranged from 2 days to 115 days with a mean of 31 (±23.6). The mean LOS of pedestrians was not significantly different from that of all other road users combined (p = 0.8339). The mean for pedestrians of 31 (±23.6) ranked third after that of motor vehicle passengers and drivers, with means of 34.17 (±26.1) and 33 (±19.8), respectively.

Overall, demographic characteristics of pedestrians admitted to a hospital, types of injuries sustained and mean length of hospital stay help us understand the burden on hospital and family resources brought about by road traffic crashes. The findings of this study indicate that males were the dominant age sub-group of admitted pedestrians. It also indicates that pedestrians comprised 62% of the total hospital bed days spent in orthopaedic wards of a major hospital in Nairobi City. Although the study did not include the charges associated with hospitalization of injured pedestrians, and although the mean

Table 5.3 Mean length of hospital stay and bed days by different categories of road users admitted, Kenyatta National Hospital, June–August 2011

Category of road user	Length of hospital stay			
	Number of cases*	Bed days, n (%**)	Mean LOS (SD ±)	P-value
Pedestrians	83	2,573 (61.6)	31.00 (±23.7)	0.9034
Motor-vehicle passengers	30	1,025 (24.5)	34.17 (±26.1)	0.4331
Motorcyclists	13	384 (9.2)	29.54 (±20.2)	0.7901
Bicyclists	6	132 (3.2)	22 (±18.7)	0.3281
Drivers	2	66 (1.58)	33 (±19.8)	0.9132
All categories	134	4,180 (100)	31.19 (±23.5)	0.8339

*Information on the mean length of hospital stay and bed days for each category of road user was calculated based on the number of road trauma admissions for the category of road user which had complete information on date of admission and date of discharge.
**Represents the percent of the total bed days for all road users.

length of stay for pedestrians was not statistically different from that of other road users admitted to the hospital, the finding of this study, that about 62% of hospital beds were occupied by pedestrians, indicate that injured pedestrians impose a huge economic burden on the Kenyan economy. Moreover, the findings of the study that the admitted pedestrians were predominantly male (male:female ratio = 3.4:1), and mainly in the productive age, 15–44 year olds (70%), underlines the huge economic and social burden caused by injured pedestrians during this productive age.

Male predominance in traffic fatalities has consistently been documented in all regions, and across all age groups, previously (Nantulya et al., 2003; Odero et al., 2003; Peden et al., 2004; Yee et al., 2006). This study adds to the evidence for the disproportionate non-fatality injury burden of the male pedestrians. Male to female ratio ranged from 1.5:1 in the age bracket of 15 to 29 to 8.4:1 in the 30 to 34 age group. Young adult pedestrians in the age group 15 to 44 are the group most frequently injured (70%). Our findings do not support the finding that pedestrian injuries are most prevalent among young children of ages 5 and 9 years, and older adults over 70 years of age, as reported in high-income countries (Vestrup and Reid, 1989; Traffic Safety Facts 2001, 2002; Retting et al., 2003).

The over-representation of males in traffic injuries, observed in a referral hospital in Nairobi, cannot be explained by population characteristics in terms of sex differentials; the estimated populations of males and females in Nairobi in the year 2009 were 1,605,230 and 1,533,139 respectively, with a male to female ratio of about 1:1.1 (Kenya National Bureau of Statistics, 2010). The male to female ratio in road traffic injury admissions was about three times that of the population in Nairobi (3.6:1 for traffic injury admissions compared to 1:1). Over-representation of males has been demonstrated in several studies (Friis and Sellers, 1996). This is probably due to the greater exposure of men to traffic, higher risk-taking behaviour among men, or increased risk due to other factors, given similar exposure levels (Odero et al., 1997; Peden et al., 2004).

Conclusion

This chapter presented findings on age and sex, the types of injuries and length of hospital stay treatment of injured pedestrians admitted to a referral hospital in Nairobi. The results show that males in the productive age category of 15–44 years were the most affected. It also shows that pedestrians occupied a significant proportion (62%) of hospital beds, thereby exerting a huge burden to the health care facility resources. Most of the injuries occurred to the limbs. Given that limbs are used as organs of locomotion, and for carrying out several key tasks for survival, the impact of the injuries to limbs might be more profound.

References

Anderson, D.R., Sweeney, D.J. and Williams, T.A. 1994. *Introduction to statistics: concepts and applications*, 3rd edition. Minneapolis, MN: West Publishing Company.

Friis, R.H. and Sellers, T.A. 1996. *Epidemiology for public health practice*. Gaithesburg, MD: Aspen Publishers, Inc.

Kenya National Bureau of Statistics. 2010. *Statistical abstract 2010*. Nairobi: Government Printer.

Mock, C.N., Arreola-Risa, C. and Quansah, R. 2003. Strengthening care for injured persons in less developed countries: a case study of Ghana and Mexico. *Injury Control and Safety Promotion*, 10: 45–51.

Mock, C.N., Quansah, R.E. and Addae-Mensah, L. 1999. Kwame Nkurumah University of Science and Technology continuing medical education course in trauma management. *Trauma Quarterly*, 14: 345–348.

Nantulya, V.M., Sleet, D.A., Reich, M.R., Rosenberg, M., Peden, M. and Waxweiler, R. 2003. Introduction: the global challenges of road traffic injuries: can we achieve equity? *Injury Control and Safety Promotion*, 10: 3–7.

Odero, W., Garner, P. and Zwi, A. 1997. Road traffic injuries in developing countries: a comprehensive review of epidemiological studies. *Trop Med Int Health*, 2(5): 445–460.

Odero, W., Khayesi, M. and Heda, P. 2003. Road traffic injuries in Kenya: magnitude, causes and status of intervention. *Injury Control and Safety Promotion*, 10: 53–61.

Ogendi, J. 2014. Characteristics of road traffic injuries to pedestrians in Nairobi, Kenya: implications for urban safety planning. PhD Thesis, Maseno University.

Ogendi, J., Odero, W., Mitullah, W. and Khayesi, M. 2013. Pattern of pedestrian injuries in the city of Nairobi: implications for urban safety planning. *Journal of Urban Health: Bulletin of the New York Academy of Medicine*, 90(5): 849–856.

Peden, M., Scurfield, R., Sleet, D., Mohan, D., Hyder, A.A., Jarawan, E., et al. 2004. *World report on road traffic injury prevention*. Geneva: World Health Organisation.

Retting, R., Ferguson, S. and McCartt, A. 2003. A review of evidence-based traffic engineering measures designed to reduce pedestrian-motor vehicle crashes. *American Journal of Public Health*, 93: 1456–1463.

Traffic Safety Facts 2001, 2002. Traffic Safety Facts 2001. Washington, DC: National Highway Traffic Safety Administration, US Department of Transportation.

Van Rooyen, M.J., Thomas, T.L., and Clem, K.J. 1999. International emergency medical services: assessment of developing pre-hospital systems abroad. *Journal of Emergency Medical Services*, 17: 691–696.

Vestrup, J.A. and Reid, J.D. 1989. A profile of urban pedestrian trauma. *Journal of Trauma*, 29: 741–745.

World Health Organisation. 2013. *Pedestrian safety: a road safety manual for decision-makers and practitioners*. Geneva: World Health Organisation.

Yee, W.Y., Cameron, P.A. and Bailey, M.J. 2006. Road traffic injuries in the elderly. *Emergency Medicine Journal*, 23(4): 42–46.

6 Safety of vulnerable road users on a road in Kinondoni municipality, Dar es Salaam, Tanzania

Estomihi Masaoe

Introduction

In low- and middle-income countries, the seriousness of the road safety risk for vulnerable road users (VRUs) has been highlighted (Peden et al., 2004). In the city of Dar es Salaam, of which Kinondoni Municipality is one of the three municipalities, VRUs constituted 79% of road fatalities during the years 2007 and 2008. Pedestrian fatalities made up 67% of the total road-based fatalities. This chapter presents findings on the VRU safety situation in the Kinondoni Municipality and the results of observation of pedestrian facilities and behaviour along Morogoro Road, one of the main radial arterial roads crossing the municipal area. The chapter particularly describes the characteristics of the road crashes involving VRUs, and the interaction between the infrastructure and the users.

Methods

The study used police records of road traffic crashes occurring in 2008 to characterize the road crashes involving VRU casualties. Morogoro Road was taken as a case site for the assessment of provision of pedestrian infrastructure along the main roads, where most of pedestrian fatalities occur. Morogoro Road is a dual carriageway from the junction with Bibi Titi Mohamed Road to Kimara Mwisho and a single carriageway thereafter. Assessment of pedestrian facilities and observations of pedestrian and driver behaviour were carried out at zebra crossings along Morogoro Road. The prevailing level of service on selected zebra crossings and sidewalks was analyzed using methods described in pedestrian and bicycle chapters of the Highway Capacity Manual (Transportation Research Board, 2000). Prevailing traffic volume and speed were compared to warrants described in the Manual on Uniform Traffic Control Devices (FHWA et al., 2003) to verify whether the use of zebra crossings are appropriate for the road.

Characteristics of vulnerable road user crashes

About 53% of all fatalities occurring in Dar es Salaam in the year 2008 were reported in Kinondoni Municipality. This section describes results of analysis of 1,180 crashes in Kinondoni Municipality involving VRUs that were reported to the police for the year 2008. The distribution of road traffic casualties by mode and gender is presented in Table 6.1. Among the vulnerable road users, the pedestrians were the leading mode, followed by motorbike riders. It is appropriate to note that along the Morogoro Road, motorcycles are observed or seen serving the feeder roads – providing transport between the bus stop and the origin or destination. During peak hours, travellers in a hurry are opting to use motorcycles instead of public transport, in spite of the higher risk involved. The use of this mode is, therefore, increasing very rapidly and so is the number of related road fatalities.

There is a high frequency of fatalities, especially males, between 18h00 and 22h00. These are the hours that tired workers are rushing home through congested roads, most of them with inadequate road lighting.

The distribution of crashes by day of week is presented in Figure 6.1, which shows that the peak was on Tuesday. Most cities around the world (see, for example, Vanderschuren and Jobanputra, 2010) find road fatalities peaking

Table 6.1 Distribution of road traffic casualties by mode and gender

Mode	Pedestrian	Bicycle	Auto rickshaw	Motorcycle	Tricycle	Total
Number of crashes	783	134	40	191	32	1,180
Fatalities	129	11	2	25	2	169
Persons injured	695	116	43	172	30	1,056
Total casualties	824	127	45	197	32	1,225
Percent casualties	67.3	10.4	3.7	16.1	2.6	100
Percent of females	17.88	0.33	0.08	0.24	0	18.53

Source: Traffic Police records for the year 2008.

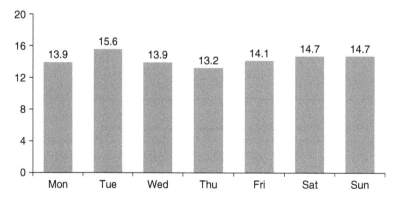

Figure 6.1 Distribution of crashes by day of week (%).
Source: Based on Traffic Police records for the year 2008.

on the weekend, due to speeding and drinking and driving. In Kinondoni Municipality, Friday through Sunday has the highest frequency, apart from Tuesday.

Table 6.2 shows that the most common action of a pedestrian at the time of crashes was crossing a road. Table 6.3 shows that contributing factors to the crashes were not known in 57% of the cases. Inappropriate speed was a contributing factor in 25% of the cases. Most of the reported cases were on the main roads (arterials and collectors).

From the analysis of crash data it is noted that the available information does not indicate the contributing factors, other than inappropriate speed. Observations of road user behaviour and the risk factors of the road infrastructure were undertaken to identify other contributing factors. The focus of the following sections is on crossing facility conditions and provision, pedestrian crossing behaviour, facility level of service and warrants compliance to provide insight into the contributing factors along Morogoro Road, the arterial road with the highest concentration of public transport services in the municipal area.

Spacing and condition of zebra crossings

The length of Morogoro Road surveyed was 14.7 km with 32 zebra crossings. On average, there was a zebra crossing after every 460 metres, which is reasonable for an urban road. On a scale of one to four (1 = pavement markings as new, visible pedestrian crossing sign in good condition and a provision for physically challenged pedestrians and 4 = pavement marking invisible to motorists or pedestrian sign is damaged/not visible to motorists), 26 zebra crossings had a score of 2 while 6 had a score of 3. This means that the pavement markings were in good condition, pedestrian signs were clearly visible but no provisions for physically challenged pedestrians were available. The general impression is that the pavement markings along this road were recently maintained and the road signs are in good condition.

Table 6.2 Action of the pedestrian victim during the crash

Action	Crossing	Standing along the road	Not reported	Total
Number	396	19	368	783
Percent	50.6	2.4	47.0	100.0

Table 6.3 Recorded contributing factors for pedestrian crashes

Factor	Not known	High speed	Careless	Over-taking	Poor road	Drunk pedestrian	All
Number	449	200	118	4	7	5	783
Percent	57.3	25.5	15.1	0.5	0.9	0.7	100.0

Pedestrian crossing behaviour

Pedestrian crossing behaviour was observed at and within 50 metres of formal crossing facilities (zebra crossings with and without pedestrian barriers, signalized intersections and pedestrian overpasses). At the Manzese section, where barriers are provided to force pedestrians to use the overpass and zebra crossings, observations showed that within 50 metres of either side of the overpass, 90.2% of pedestrians used the bridge while 9.8% crossed at between 20 metres and 50 metres away from the overpass, at a point where the fence was damaged (n = 2,705 during a 3.5-hour period). Likewise, observation of a particular zebra crossing (near Bakhresa bus stop) revealed that 86.9% of observed pedestrians used the crossing facility while 10.2% crossed between 20 metres and 50 metres to the left of the facility, while 2.9% crossed between 20 metres and 50 metres to the right of the facility, where the barrier was damaged (n = 8,288 observed within 2 hours around mid-day). A negligible number of pedestrians (less than five pedestrians) crossed between the crossing and a distance of 20 metres on either side of the crossing. The Manzese area has a very high concentration of pedestrians crossing the road due to commercial activities on either side of the road. The use of pedestrian barriers effectively channels pedestrians to specific crossing points. However, the need to maintain the barrier is apparent.

Pedestrian crossing behaviour at locations without barriers seemed to be influenced more by their desire lines/destination than the number of formal crossing facilities. Near Kibo/Kona bus stop only 54.5% crossed at the formal crossing area (Figure 6.2).

Observations at signalized intersections, with properly functioning pedestrian signals, revealed that pedestrians generally did not follow the signal indication and fairly few crossed at the formal crossing facility, as shown in Figure 6.3. This behaviour may be partially attributed to the practice of police officers overriding signal control, most of the time, and their habit of totally ignoring the needs of pedestrians, as reported by Nyoni and Masaoe (2011). Pedestrians are left to determine on their own where and when they perceive it to be safe to cross. Those with destinations further

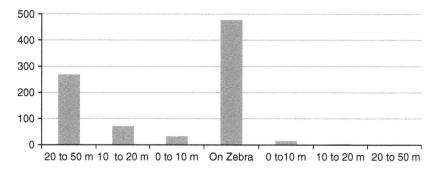

Figure 6.2 Pedestrian crossing behaviour at zebra crossing (3h observation, n = 868).

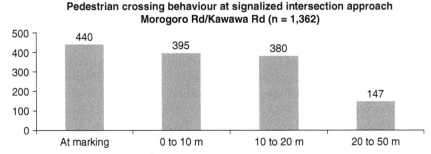

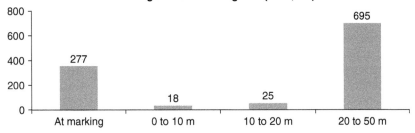

Figure 6.3 Pedestrian crossing behaviour at signalized intersections (Elibariki, 2011).

away from the formal crossing conveniently cross more than 20 metres away from the markings.

Thus, a decision to install pelican crossings, in an environment where traffic signals are routinely overridden by traffic police, should be carefully evaluated. Soft separation by use of raised zebra crossings may be a better option, especially if they are located to coincide with the pedestrians' desire lines and spaced so as to manage vehicle speeds to the desirable maximum of 50 km/hr, to assure safety of the pedestrians and vehicle throughput. This approach is similar to the application of traffic calming measures on rural arterial roads passing through a settlement, where the mobility function of the road is curtailed so as to achieve safety and environmental goals (known as an 'environmentally adapted arterial road' according to Ogden, 1996). This approach is sensible, given that the great majority of the trips are on foot and by bus, and managing the road transport system to favour the minority of private car trips is unacceptable. However, reconstructing the road to accommodate bus rapid transit and other options may be more appropriate, but what is proposed is applicable to all roads with high pedestrian and vehicle demand, which makes zebra crossings ineffective as the only provision for pedestrians to cross. This proposal is in agreement with the findings by de Langen et al. (2004) and seems to have been partially implemented on the section of the road between Ubungo and Mbezi Mwisho. In this section, road humps were installed instead of the recommended raised zebra crossings with a flat top, using bricks with approach slopes selected to achieve the desired maximum vehicle speed. The economic advantages of raised zebra crossings (trapezoidal section constructed using paving blocks), in terms of lower maintenance costs and lower vehicle operating costs are also advanced according to de Langen et al. (2004). De Langen et al. also noted that driver route selection in Nairobi did not seem to be affected by their presence. A more systematic application of raised zebra crossings throughout the road where the 85th percentile speed exceeds 50 km/h is desirable to assure safe mobility of pedestrians and other VRUs.

Observations at zebra crossings adjacent to bus stops revealed that, when the crossing facility provided an apparently safe and logical link from the bus stop to the access road on the other side of the road, its usage was high; otherwise usage tended to be low. For example, at Kimara Mwisho, where the zebra crossing location was counter-intuitive to the path a pedestrian would choose when walking to or from the bus stop (for passengers going to the city centre), only 20% used the crossing facility, compared to those crossing within 50 metres of the crossing facility. At Mbezi Mwisho the behaviour is very different with 85% using the crossing. This is because the pedestrians are assisted by traffic police, since traffic is so heavy and intervention by an enforcement officer is necessary to create gaps for pedestrians to cross. Between Kimara Mwisho and Mbezi Mwisho at Kimara Stop-Over, a raised zebra crossing was used by over 90% of the crossing pedestrians, since the raised crossing helped to create gaps in the heavy traffic. At this location it seemed as if a raised zebra crossing could function like a pelican crossing, minus the unnecessary delays to vehicular traffic. This phenomenon will be examined quantitatively in a future work.

Overall, the results of observation of pedestrian crossing behaviour shows that pedestrians do not necessarily prefer to use zebra crossings or other crossing facilities. Behrens (2010) reported similar observations for the main roads in Cape Town, where he noted that crossing facilities with high utilization rates tend to coincide with the pedestrian desire lines. A study on effectiveness of application of raised zebra crossings in Dar es Salaam (de Langen and Tembele, 2001; de Langen et al., 2004) noted that provision of raised zebra crossings does not cause a change in the pedestrian crossing patterns, but improves safety through the general reduction of speed along its area of influence. To encourage pedestrians to use formal crossings, the use of guiding barriers seems to be necessary. This was successfully implemented along the section passing through Manzese, where there is a very high volume of pedestrians crossing from one side of the road to the other. There was a very low utilization rate of the overpass until barriers were used to compel crossing pedestrians to use it.

Level of service of crossing facilities

Level of service is an indicator of the quality of service experienced when using a traffic facility. This is indicated by the average waiting time or delay road users experience, in this case pedestrians when they wish to cross at a particular location. Three zebra crossings were selected to determine the level of service. Waiting times and crossing speed of randomly selected pedestrians (30 males and 30 females) were determined. Figure 6.4 presents results for the average delay. According to the Highway Capacity Manual (Transportation Research Board, 2000) the levels of service at the Manzese, Kimara Mwisho and Mbezi Mwisho crossings were D, D and E, respectively (the Mbezi Mwisho crossing was police assisted).

Appropriateness of crossing facilities

The question the author addressed was whether the zebra crossing facilities are appropriate for the road, given its traffic volume and vehicular speed. Vehicle volume and speed were measured at three locations and compared to the warrants recommended in FHWA et al., (2003). Figure 6.5 presents the results. It is clear that the current demand is beyond the recommended maximum hourly pedestrian volumes for a zebra crossing for all three crossings.

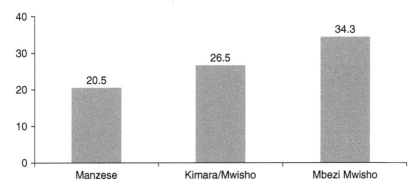

Figure 6.4 Average delay and crossing time at three zebra crossings (Rajabu, 2011).

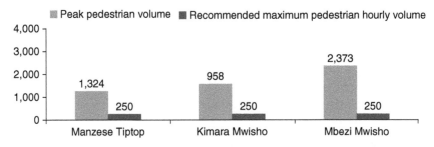

Figure 6.5 Pedestrian hourly volumes compared to recommended maximum (Rajabu, 2011).

There is a need for the managers of the infrastructure to re-evaluate their disproportionate provision for the motorized traffic at the expense of pedestrian traffic.

Figure 6.6 presents a comparison of vehicle volume and the recommended maximum hourly volume for zebra crossings for three sections of the road. Clearly the vehicle traffic demand means that zebra crossings are not appropriate crossing facilities. The results of average speeds estimation at the Manzese, Kimara and Mbezi sections were 38 km/h, 56 km/h and 51 km/h, compared to posted speed limits of 40 km/h, 50 km/h and 50 km/h, respectively. The 85th percentile speeds exceeded the posted speed limit, which shows that the behaviour of many of the drivers did not comply with the speed limits.

The recommended vehicle and pedestrian volumes for application of zebra crossings, for example, by FHWA et al. (2003), suggest that the demand of both pedestrian and vehicle traffic is too high for zebra crossings to be a safe, convenient and efficient infrastructure measure. The low level of service at the selected crossing facilities confirms this.

Driver–pedestrian interaction at zebra crossings

Behaviour of vehicle drivers approaching a zebra crossing was observed 30 metres before the zebra crossing. The purpose was to note how they interacted with pedestrians wishing to cross at the zebra. It was found that only 10% of drivers yielded/stopped to allow pedestrians to cross.

The sidewalk

Between Magomeni and Ubungo the service roads serve as sidewalks although, with few exceptions, bicycle and vehicular traffic mixes with pedestrians. Between Ubungo and Kimara/Butcher the service roads are wider and

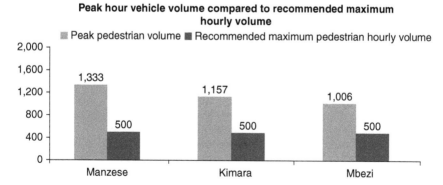

Figure 6.6 Vehicle volume and recommended volume for zebra crossings.

pedestrian and vehicle volumes are, generally, low. Between Kimara/Butcher and Kimara Mwisho the service roads have a gravel surface, which is not attractive for pedestrians. Beyond Kimara Mwisho, where the road cross-section changes from dual to single carriageway, there is no sidewalk or service road for most of the road length and pedestrians have to walk in the shoulders. Observations of pedestrians walking on the service road/sidewalk were carried out at Ubungo near the upcountry bus terminal and at Manzese. There are business activities being carried out at this section, which narrows down the service road and increases interactions between the pedestrians and those involved in commercial transactions. Observations of speed of pedestrians and bicycles, and their interactions per hour, compared with the Highway Capacity Manual (Transportation Research Board, 2000), indicated that the sections at Ubungo and Manzese, during the peak hour, were operating at level of service D and C, respectively. The overall assessment was that the presence of traders on the sidewalk decreases the level of service of the facility substantially.

Conclusion

This chapter shows that pedestrian crossing behaviour in terms of utilization of formal crossings varies and seems to be influenced by presence of barriers and the location of the crossing facility, relative to their preferred path. The zebra crossing facilities along Morogoro Road do not conform to the warrants (FHWA et al., 2003). Thus, that provision of zebra crossings along the road, given its high pedestrian crossing demand and vehicle traffic speed and volume, does not offer sufficient safety and convenience to the pedestrians.

References

Behrens, R. 2010. Pedestrian arterial and freeway crossing behaviour in Cape Town: observations and implications. 12th WCTR, 11–15 July 2010, Lisbon.
de Langen, M. and Tembele, R. 2001. Guidelines for pedestrian and bicycle traffic in African cities. IHE-Delft, Balkema (eds), prepared under the WB-SSATP.
de Langen, M., Rwebangira, T., Kitandu, E. and Mburu, S. 2004. Urban road design in Africa: the role of traffic calming facilities. Available at: www.gtkp.com/assets/uploads/20091130-104714-8951-delangen.pdf (accessed on 12 August 2011).
Elibariki, C. 2011. Pedestrian safety and behaviour in the city of Dar es Salaam. MSc Dissertation, Department of Transportation and Geotechnical Engineering, University of Dar es Salaam.
FHWA, ATSSA, AASHTO and ITE. 2003. *Manual on uniform traffic control devices for streets and highways* – 2003 edition. Washington, DC: FHW.
Nyoni, J. and Masaoe, E. 2011. Factors contributing to high frequency of vulnerable road user injury in Dar es Salaam. Proceedings: the 30th SATC, 11 to 14 July 2011, Pretoria.
Ogden, K.W. 1996. *Safer roads: a guide to road safety engineering.* Avebury Technical, UK.

Peden et al. 2004. World report on prevention of road traffic injury, Geneva: WHO.

Rajabu, M. 2011. Evaluation of Pedestrian Facilities along Morogoro Rd, Dar es Salaam. Unpublished Project Report, Department of Transportation and Geotechnical Engineering, University of Dar es Salaam.

Transportation Research Board. 2000. *TRB special report 209: the highway capacity manual (HCM2000)*. Washington, DC: TRB.

7 Non-motorized transport infrastructure provision on selected roads in Nairobi

Winnie V. Mitullah and Romanus Opiyo

Introduction

Provision of infrastructure for non-motorized transport (NMT) lags behind in most African cities, as reflected in the absence of NMT policy, legislation, standards and guidelines in most cities (see Chapters 3 and 12 in this book). In Kenya, despite formulation of Sessional Paper No. 2 of 2012 on integrated national transport policy, NMT is still not embedded in overall planning and infrastructure development of cities. It is provided along motorized routes, depending on availability of space, and often without consideration of the principles of safety, coherence, directness, attractiveness and comfort. Furthermore, the design standard of infrastructure is not uniform and continuous, and does not take into consideration the origin and destination of NMT users. These shortcomings result in inefficient use of facilities and conflict with motorized transport and enforcement officers.

In Nairobi, focused NMT infrastructure provision dates back to a mapping and situational analysis of NMT infrastructure along 18 routes, in 2006, as part of a study on a master plan for urban transport in the Nairobi metropolitan area, popularly known as NUTRANS (Katahira and Engineers International, 2006). Since then, provision of NMT infrastructure has become mandatory for new roads, while older roads are retrofitted. The majority of Nairobi residents, in particular low-income households, use NMT as their primary mode. This notwithstanding, NMT infrastructure has not been prioritized. However, this is expected to change if the Nairobi NMT draft policy, developed in March 2015, is followed by legislation and related guidelines for NMT development.

This chapter examines the availability of NMT infrastructure and facilities in 18 major road corridors in the City County of Nairobi (CCN), and the adequateness of the infrastructure. It is based on mapping the situation analysis of the 18 routes[1] in 2006, 2011 and 2015 and a literature review on NMT travel analysis and infrastructure provision by the Institute for Development Studies (IDS) of the University of Nairobi. The 2011 and 2015 mapping was aimed at updating an initial inventory of NMT infrastructure, undertaken by Katahira and Engineers International (2006) and assessing available infrastructure and facilities in line with the principles for NMT development. Some of the proposals by Katahira and Engineers International have since been addressed and are discussed in this chapter.

Methods

In carrying out the NMT infrastructure mapping and analysis, the research process used a combination of techniques of information gathering, using the Katahira and Engineers International study (2006) as a baseline. The exercise was conducted in three phases: pre-field, field and analysis. During the pre-field phase the researchers prepared 19 base maps for routes proposed for development of NMT infrastructure (see Figure 7.1). The preparation included a review of the satellite imageries from various sources, such as Google Earth and Wikimapia, in order to establish the gaps in spatial data. In addition, a review of data sources within the government, including the Kenya Roads Board (KRB), was done to establish the roads network in Nairobi and to assist in further classification. Equipped with this knowledge, digitization of maps followed, using the already acquired topo-sheet and satellite imageries. These were digitized using GIS and merged to get a glimpse of the Nairobi roads network. This output was used as a guide for the field survey and assisted in coming up with an updated thematic map of NMT infrastructure along the 18 routes.

During the same phase, an observation checklist and method of data collection was developed and agreed upon. The following issues were isolated for observation, namely, walkways, NMT bridges, pedestrian crossings, existence of dedicated cycle tracks and intersection designs. In examining walkways, researchers assessed issues such as separation from motorized

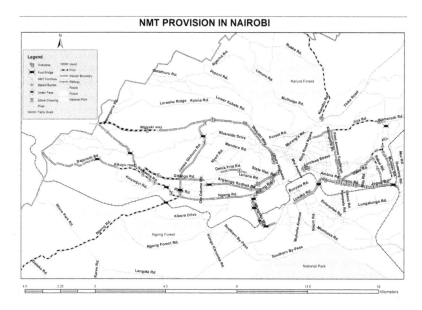

Figure 7.1 Nairobi roads NMT provisions.
Source: Field Survey, 2015.

vehicles, dedication of infrastructure to walking, prevention of motor vehicle intrusion, paved infrastructure without obstructions and safe crossings. Basic assessment criteria were used, namely 'good footpath' (indicating new and well-maintained paved footpath), 'fairly good footpath' (indicating not new, unkept and showing signs of worn out paving), 'poor condition footpath' (indicating not paved) and 'non-existent' (inferring no footpath).

In assessing safe and convenient pedestrian crossings, 'good' indicated raised and painted zebra crossings, islands for pedestrians and cyclists, and 'bad/faint marking' indicated not visible but existing. For footbridges, 'paved' meant well-constructed and cemented to ease passage of pedestrians, and 'unpaved' meant not cemented and difficult to use during the rainy seasons (Table 7.1). The study also assessed the existence of dedicated bicycle tracks, and appropriate intersection designs, which enable efficient NMT traffic flow.

Table 7.1 NMT facility rating criteria

NMT facility	*Indicator*	*Measurement scale*
1. Footpath/cycle path	Excellent	Outstanding accessibility, available on both sides of the road, buffer zone between vehicular and pedestrian traffic, paved and intact surfaces, outstanding and complete connectivity with other modal facilities
	Good	Paved and intact surfaces, buffer zone between vehicular and pedestrian traffic
	Fair	Existence of paved surfaces
	Poor	Dusty or poorly maintained facility
2. Pedestrian crossing and speed bump	Excellent	Visible zebra markings, raised crossings, strategically located road sign with signal control
	Good	Visible zebra markings
	Fair	Faded zebra markings, no signs indicating existence of the facility
3. Overpass and footbridge	Excellent	Paved and intact surface, existence of continuous hand rails to help people in danger of slipping and falling, appropriate ramps with smooth ground level access for people with disabilities and strategically located
	Good	Paved and intact surface, existence of hand rails, existence of ramps and strategically located
	Poor	Dusty or poorly maintained facility, without ramps and hand rails

During the field work the research team used GPS to take coordinates of the location of various facilities. The team assessed and recorded the location of the NMT infrastructure and their existing condition. This was collated using photographs of the facilities observing the behaviour of the users and conflicts between users.

The analysis and interpretation stage included processing the spatial data collected. The geo-referenced points collected were transferred to the Geographic Information System (GIS) and digitized. This showed the specific location and distribution of NMT facilities in various routes. Production of updated base maps show NMT facilities as the main thematic issue of Nairobi NMTs. The photographs taken were used to show the condition of the NMT facilities, areas of observed conflicts and users of the NMT facilities. The main focus of analysis was to find any emerging unique feature of the NMT condition in 2011 and 2015.

In analysing the information, the study used the principles of safety, coherence, directness, attractiveness and comfort (see Table 7.2).[2] However, the study used only four variables because the users were not interviewed on their origin and destination. This made it impossible to conclusively measure directness in both 2011 and 2015.

Although the inventory and analysis in this chapter may not be comprehensive, it provides useful information for understanding NMT infrastructure provision in Nairobi and may be used in future assessment of progress.

Table 7.2 NMT infrastructure design and provision principles

Principle	*Explanation*
Safety	Maximization of the safety of NMT users, in relation to other road users, since they have a high degree of vulnerability
Coherence	Formation of a coherent and continuous network linking all origin and destination points for NMT users, and not ad hoc facilities that end abruptly
Directness	Extent of formation of a direct route from origin to destination without significant detouring that is likely to cause the users to ignore the facility
Attractiveness	Ability of NMT facilities to make NMT travel attractive, both by day and night
Comfort	Guarantee a smooth, quick and comfortable flow of NMT traffic without excessive gradients or uneven surfacing

Source: translated from Ploeger, 1993.

Status of NMT infrastructure provision on the 18 road corridors

The routes were categorized into Zone A (Westlands Zone) – covering all roads on the upper side of Uhuru highway – and Zone B (Eastlands Zone) – covering all roads below Uhuru highway, as shown in Table 7.3.

Zone A roads

Ngong Road

This road connects Ngong Town and Nairobi's Upper Hill Area (see Figure 7.2). It is estimated to be 28 km from Nairobi CBD. The length of the paved foot-path was estimated to be 6.5 km, which was about 23% of the total length.

There is no provision for cyclists along this route. In terms of NMT, it serves as a link route to parts of low-income areas (Kibera and Kawangware), and also connects to major activity nodes around Kenyatta Hospital, the Upper Hill Community, Adams Arcade and Dagoretti Corner. These nodes are characterized with dense NMT activities, hence, linking a number of NMT users, mainly pedestrians and cyclists. There is provision of paved footpaths from the Nakumatt Junction area towards City Mortuary. However, the footpath is disconnected at some points, and some parts covering Dagoretti Corner and the Ngong area through Karen are unpaved.

The Naivasha–Ngong Road Junction marks the beginning and the end of paved and unpaved footpaths in both directions. It is the starting point of the paved footpath towards the City Mortuary and also marks the beginning of the unpaved footpath towards the Karen area. There are few well-marked and visible zebra crossings along this road. Around Kenya Science Campus there are speed bumps that are 35.3 metres apart with a pedestrian crossing in between; however, there is no signage to indicate any speed bumps or zebra crossing ahead. The speed bumps are only noticeable at close range.

Some peculiar behaviour among NMT users was noted around Impala Sports Ground, where pedestrians walk away from the paved pedestrian strips. They prefer using the unpaved path, especially during dry seasons. This is because the paved section is so close to the arterial road, and risky to use. It was also noted that the cyclists along this route always ride on the arterial road, since there is no separate provision for them. Figure 7.2 shows some of the noted features of NMT along the Ngong Road, including streetlights around the Adams Arcade area which, in terms of safety, makes it possible for NMT users to use the facility, even at night. Storm drains are in poor condition, and are hardly noticeable as grass has overgrown them and they are not at all functional during the rainy season. This causes inconvenience to pedestrians as the footpaths get muddy and flooded. Around the junction to Kibera, next to Nakumatt Prestige Shopping Mall, traders have encroached on the footpaths forcing the pedestrians to use the road edge as a path, which is also used by the *matatus* (i.e. informal public transport) as a pick-up and drop-off point for passengers.

Table 7.3 Nairobi NMT analysis summary table

Road name and zone	NMT facility type availability (yes/no)									Total length in km
	Footpath	Bicycle path	Pedestrian crossing	Speed bumps	Footbridge/ underpass	Traffic lights	Other NMT support furniture	Condition		
A. Westlands Zone										
1. Ngong	Yes	No	Yes	No	No	No	No	Fair		9.6
2. Kikuyu	Yes	No	Yes	Yes	Yes	No	No	Fair		6
3. Naivasha	Yes	No	Yes	Yes	Yes	No	No	Fair		6.3
4. James Gichuru	Yes	Yes	Yes	Yes	Yes	No	No	Fair		4.6
5. King'ara	Yes	No	Yes	Yes	Yes	No	No	Good		1.2
6. Gitanga	Yes	Yes	Yes	Yes	No	No	No	Good		2.7
7a. Argwings Kodhek	Yes	No	Yes	No	No	No	No	Fair		3.6
7b. Argwings Kodhek Extension	Yes	No	Yes	No	Yes	No	No	Fair		0.5
8. Dennis Pritt	Yes	No	Yes	Yes	No	No	No	Fair		2.0

Table 7.3 Nairobi NMT analysis summary table

Road name and zone	NMT facility type availability (yes/no)							Condition	Total length in km
	Footpath	Bicycle path	Pedestrian crossing	Speed bumps	Footbridge/ underpass	Traffic lights	Other NMT support furniture		
9. Mbagathi Way	Yes	No	No	Yes	Yes	Yes	No	Good	3.0
10. Valley Road	Yes	No	Yes	No	Yes	No	Yes	Good	2.3
11. Waiyaki Way	Yes	No	Yes	Yes	No	No	Yes	Fair	3.0
B. Eastlands Zone									
12. Kiambu	Yes	No	Yes	Yes	Yes	No	No	Fair	5.6
13. First Avenue Eastleigh	Yes	No	Yes	Yes	Yes	No	No	Fair	3.2
14. Heshima'	Yes	No	No	Yes	No	No	No	Fair	4.2
15. Mumias	Yes	No	Yes	No	No	No	No	Fair	5.9
16. Rabai	Yes	No	Yes	Yes	No	Yes	No	Fair	1.8
17. Jogoo	Yes	No	Yes	No	Yes	Yes	No	Fair	5.2
18. Lusaka	Yes	No	Yes	No	Yes	No	No	Fair	3.5

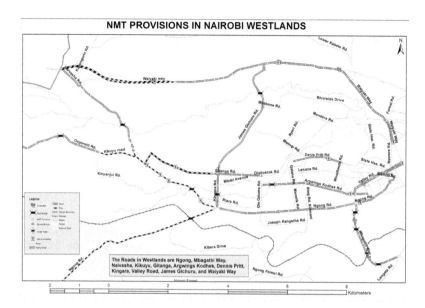

Figure 7.2 NMT provision on roads in the Westlands.
Source: Field Survey, 2015.

In terms of general safety for users, in relation to other modes, the buffer mechanism is poor. The facility on this route is prone to conflict between NMTs and motorized transport, on the one hand, and on the other, between different modes of NMTs, for example, between cyclists and pedestrians. This is because of a lack of provision for cyclists. In terms of lighting, there is a good attempt to provide lighting in some sections of the route, albeit inadequate.

The NMT network provided is also not coherent and attractive, as the network is not continuous, forcing users to leave the facility at certain points, for example, at the junction near Postal Offices past the Rugby Football Union of East Africa (RFUEA) grounds. There is no visible expansion and segregation of various NMT users. Discontinuity of the provided NMT facilities assumes that NMT is not solely used as a mode, but a mode to take a bus and other means to different destinations, which is misleading. It is, however, not possible to address the directness of the NMT users' routes without the origin-destination data of NMT users. The footpaths/walkways are less than 2 m wide and, given the current moderate pedestrian flows, most people are forced to walk far away from the designated facility, which is not safe, given the frequent invasion of walkways by speeding *matatus* which tend to overlap during traffic snarl-ups.

Kikuyu Road

Kikuyu Road links Riruta to Ruthimitu, Waithaka, Thogoto and Kikuyu Town, a distance of about 25 km. It attracts NMT modes around Riruta

Satellite, Waithaka Shopping Centre and the Thogoto area, which hosts various institutions, such as the Alliance Schools and the famous Eye Clinic and Kikuyu Town, which is a hub of many activities. The length of the paved footpath was found to be approximately 6 km, which is around 25% of the total road length. Kikuyu Road has a paved footpath, whose condition has since deteriorated, as some stretches are paved, while others are not. The footpath is narrow and cannot accommodate users adequately. Provision for other categories of users, such as cyclists, is non-existent. The traffic calming measures, such as zebra crossings, were not marked and difficult to locate and map, even in cases where they are planned.

Figure 7.2 provides the distribution of NMT facilities along the corridor. Noticeable NMT facilities available on the Kikuyu route are a footpath, which is paved and generally in good condition, and an unpaved footbridge. The footpath is, however, narrow and only meant for pedestrians. Some parts of the footpath are in bad condition, especially around the Waithaka and Ruthimitu areas. At Waithaka Shopping Centre, the footpaths have been encroached by traders and *matatus* as there are no designated pick-up or drop-off points for passengers. The narrowness of the footpath also makes it unattractive to users, especially during rainy season when all pedestrians are forced to scramble for the existing paved narrow space, to avoid mud and flooded sections of the path.

The road has insufficient streetlights, compromising security of pedestrians in the evening. There is no signage and road marking. There are also few speed bumps in the area around Feed the Children. At Dagoretti High School Junction, the footbridge is undergoing reconstruction. Provision of NMT infrastructure has not changed drastically along this route, as the provided NMT paths are not fully separated or protected, and the existing traffic calming facilities (speed bumps) are not properly maintained and there are no proper drainage facilities. However, there are some improvements, including protection of the footpath from motorists, which reduces conflict between motorists and NMTs. There are also some street light masts, which make the road safe to use at night. The safety is also enhanced by the barrier protecting NMTs from motorists, which separate pedestrians from the speeding motorized traffic on the roadway, but barriers are not continuous, and limited to where NMT infrastructure, such as cross bridges, is provided. Overall, the terrain between Waithaka and Naivasha Road Junction is fairly steep, and hence not friendly for cycling, but the distance is short (3 km), which is good for walking.

Naivasha Road

The road connects the Uthiru, Riruta Satellite and Kawangware areas to Dagoretti Corner, which is a distance of around 9 km, with a length of the paved footpath estimated to be 6.3 km (Figure 7.2). There is a continuous paved pedestrian path along the road with facilities such as zebra crossings, speed bumps and a footbridge. Consequently, the availability of NMT facilities was classified as good. The part between Kawangware and Ngong Road, which is around 3.2 km, was found to have a fair footpath and pedestrian crossing and good speed bumps. The section between Kawangware and the Institute of Livestock Research Institute (ILRI), covering a distance of 3.3 km, has a footbridge, good footpath, fair speed bumps and improvement in provision of streetlights and road signage.

Although the road had several traffic calming measures, there was no clear separation and protection of the pedestrian users, which compromised the safety of the users. Furthermore, users' comfort is not guaranteed due to cases of parking of motorcars on the facility. Opposite the Agriculture Fisheries and Food Authority (AFFA) offices, there is also encroachment on the footpath by traders; the drop-off point also seems very disorganized and the drainage is poor and cannot effectively function during the rainy season. In spite of these challenges, the footpaths were fairly coherent, with a relatively continuous network, and the footpath is used as there is no alternative.

Kingara Road

The whole footpath along Kingara Road is paved and is estimated to be 1 kilometre in length. The road attracts NMT traffic, mainly walkers from those who are living in Kawangware and Kibera and working in the Riara neighbourhood. There is also provision of traffic calming measures, notably a speed bump near Gitanga Junction, and also a narrowly paved footbridge crossing over the Kirichwa Kubwa River. There is a well-marked zebra crossing next to the Ngong Road Junction next to Nakumatt.

There is fair coherence in provision of footpaths, with some areas narrowly paved at the Kirichwa Kubwa River footbridge, while others are widely paved, such as areas next to the Gitanga Road and Ngong Road Junctions. However, the footpaths are not protected for NMT use. The general slope of the route is uneven, contributing to users' discomfort, and it is not very ideal for cycling.

James Gichuru Road

James Gichuru to Waiyaki Way is about 5 kms. It attracts NMT traffic from Kawangware to the Westlands area, and is also popular to those living in Kangemi, connecting from Waiyaki Way to Muthangari and Loreto and Strathmore School, among others. James Gichuru has a paved footpath in good condition for pedestrians covering the entire road, and the area around the Olengruone Avenue roundabout has some cycle tracks. The part between Waiyaki Way and Loreto Msongari School has footpaths, speed bumps, a pedestrian crossing and footbridge.

The most notable unpaved footpath was around Loreto Convent Msongari School, Muthangari Police Station and the Strathmore School Junction, while Waiyaki Way to Olenguruone Roundabout has a well-paved and spacious footpath, and the storm drains were very well maintained. Olenguruone Avenue has sufficient pedestrian crossings with protective bollards. There was also provision for cycle lanes with proper signage, as well as streetlights and CCTV cameras. At Isaac Gathanju Road, there is a pedestrian crossing, as well as a speed bump. Footpaths were found wanting on some stretches, but good at Kenya Dentistry, on one side of the road.

There is an attempt to retrofit NMT facilities on this route, as evidenced by the improvement from the Waiyaki Way Junction to Olenguruone Avenue Roundabout. There is no coherence in provision of footpaths; some areas are paved, while others are unpaved, and hence not fully comfortable and attractive to NMT users. James Gichuru was also found to be one of the few roads which have provision for cyclists, including crossing paths for cyclists, which are additional to what existed in 2011.

Gitanga Road

Gitanga Road is a busy road connecting NMT traffic from the low-income areas of Kawangware and Riruta Satellite to various work destinations. The area connecting Kawangware to Ole Odume Road has a paved footpath measuring around 2.7 km, with provision of speed bumps and a zebra crossing with faded paint. The section between Ole Odume Road and James Gichuru Junction is in good condition, paved, fairly attractive and comfortable. There is a cycling lane, it is not segregated, but shows the best characteristics. Some of the section was unpaved and also blocked by solid waste and parked cars, forcing pedestrians to use the carriageway. The route attracts heavy NMT traffic, mostly originating from the low-income areas of Kawangware and other neighbouring estates, such as Riruta Satellite.

From Othaya Road Junction, the footpaths are unpaved, dusty, narrow and close to the arterial road. At Maria Immaculata Hospital, there is signage for a zebra crossing but no marking to accompany it. Speed bumps are well distributed from Amboseli Road Junction all the way to Rusinga School, albeit unmarked. Protected footpaths around Amboseli Road Junction are very narrow and dangerously close to the carriageway. There is massive encroachment on the unpaved and dusty footpaths by traders and public vehicles around the Congo area.

The NMT facilities, provided in this route, are fairly safe in terms of provision of dedicated footpaths for pedestrians, though not fully paved and coherent. Provision of traffic calming measures, such as speed bumps and zebra crossings, helps with calming of motorcars to enable NMT users to maximize utilization of existing facilities. Streetlights accompanying the footpath facility are available to enhance users' security and safety during the night, although residents noted that they were not working, making the area a hot spot for mugging, especially in the evening.

Argwings Kodhek Road

Argwings Kodhek Road connects the Valley Road area with the Hurlingham and Kilimani areas. These areas are undergoing rapid densification and mixed use development, changing the neighbourhood character and population profile by bringing in a fairly young middle class population. This road traverses two major activity nodes, namely, Hurlingham Shopping Centre and Yaya Shopping Centre. Vehicles destined to Kibera and Kawangware also use the route, hence attracting a volume of NMT users who work within the catchment of this road. The length of the road connecting King'ara Road to Valley Road is around 7 km. The area connecting the Argwings Kodhek Road section, located between Kayahwe Road Junction and Woodlands Road Junction, has an approximate 2.6 km length of road, which is paved and in good condition, while approximately 1.6 km is paved and in fairly good condition. The part near Ole Odume Road Junction has a footbridge in good condition.

A sizeable length of the route has a paved footpath which also has bollards, protecting and separating the arterial road from encroaching onto the footpath. This enhances safety of pedestrians, which also makes it attractive and fairly comfortable to use. In terms of coherence, the route can be said to be fairly coherent since sections are not fully connected and there is no protection, as shown in the area around Ole Odume Road Junction.

Argwings Kodhek Road Extension

The extension of Argwings Kodhek is the section connecting Valley Road to Ralph Bunche Road. The route has a slightly paved footpath, near the junction of the Valley Road and Ralph Bunche, but most of the portion, approximately 60%, is dusty and unpaved. Another notable facility is the overpass connecting the two roads at Valley Road Junction. Other notable features were the presence of streetlights and CCTV cameras, which further enhances pedestrian security. Storm drains were also noted, but not well maintained and reliable during rainy seasons. The zebra crossings along Silver Springs Hotel and Doctors Plaza Nairobi Hospital, through to Lee Funeral Home, have completely faded and are no longer visible.

Dennis Pritt Road

Dennis Pritt is a road connecting State House Road to the Kilimani neighbourhood, which is mainly a high-income residential neighbourhood. The Kenya National Examination Council (KNEC) offices, St. George's Primary and Girls Secondary School and the entrance of the State House are all located along this road and there is a lot of NMT traffic from the bus stop to and from their destinations as the road is not served by any public service vehicle.

NMT provision on the entire length of the road is good, and the area towards Kenyatta Avenue, approximately 2 km in length, is paved. A sizeable portion of the road towards Kayaweh Road Junction area is unpaved. The route has provisions for pedestrian crossings and speed bumps; there is signage to indicate children crossing, but no zebra crossing to accompany it, at St. George's Primary School. Around Rose Dale Gardens, reconstruction of the footpath and the storm drains are underway, though the footpath along Kenya National Examination Council (KNEC) is good and very well paved, with some stretches complete with bollards.

The route has speed bumps and a pedestrian crossing near St. George's Secondary School and near the State House. The most visible NMT support facilities on this route are speed bumps and pedestrian crossings near institutions. Most parts of the route had unpaved footpaths. It was further observed that the general construction of buildings have adhered to the building lines, hence there is ample space for public utilities which is advantageous to provision of NMT facilities.

In terms of safety, most sections of the road are not safe, due to the construction activities; however, it is expected to be attractive and comfortable for NMT users once the construction is complete. The short paved footpath provided is incoherent and cannot connect complete travel points in the area. In terms of general connectivity, without consideration of the condition of roads, the section towards Kenyatta Avenue connects appropriately to Valley Road and bus stages around the National Social Security Fund (NSSF) and Serena Hotel area.

Dennis Pritt Road is one of the routes which needs proper and continuous provision of NMT facilities, especially pedestrian and cyclist paths, since there are no public service vehicles operating along this road, creating demand for NMT as a possible origin to destination mode, since those in NMT modes do not have the luxury of changing their mode at any transfer point along this route. There is notable expansion of the length of NMT coverage, as compared to 2011, with provision of footpaths on the two sides of the road and good management of existing facilities, such as zebra crossings.

Mbagathi Way

Mbagathi Way connects traffic from Ngong Road to Lang'ata Road traversing through Kenyatta Market and Nyayo Highrise Estate. It is a popular NMT route for those coming from Kibera and working in the industrial area and southern part of the city. It is one of the routes which was noted to be fairly friendly to NMT users, especially the pedestrians, as the route has a number of provisions facilitating pedestrianization. The route has a high NMT traffic flow in the morning and evening hours, mostly heading to or from Kibera. It is a highway to the industrial part of the city and many residents of the sprawling Kibera informal settlement, and Nyayo residential area, use the route.

Mbagathi has 3 kms of paved road in good condition, a footbridge, two functional overpasses and speed bumps. It has a coherent and spacious footpath connecting Ngong Road and Lang'ata Road. The width of the footpaths was estimated to be around 2 metres wide. There is also provision of raised and paved footbridges for pedestrians which is lacking in most of the other routes. It has other functional facilities, such as two paved and raised footbridges and two overpasses strategically located near high NMT traffic crossing points, which are all strategically located as crossing points connecting major land uses and destinations across Mbagathi Way, one near Kenyatta Market and Ngumo Junction and another one near Riara University Campus and Highrise Estate, which is also the crossing point to those whose destination is the densely populated Kibera and its environs.

The underpass is, however, not well designed, especially at Mbagathi, since a railway line passes right through it, which is very dangerous. Furthermore, one can easily be mugged without people on the ground being aware of such attacks. It is littered with garbage and cannot accommodate many pedestrians at the same time. On the other hand, the footbridge around Mbagathi is fairly well designed and can accommodate a disabled person on a wheelchair, due to the ramps, but the one at Highrise Estate does not have ramps; hence, the disabled cannot use it effectively. Other notable features were sufficient streetlights and CCTV cameras. The storm drainage facilities were also very well maintained, thus minimizing flooding of the footpaths and roads during the rainy season. There is also a provision of pick-up and drop-off points for *matatus*, which are very convenient, and also minimizes encroachment of footpaths by *matatus*.

The route was fairly safe, given that there is a lack of adequate buffer protecting pedestrians from motorists. Coherence of the route was achieved for pedestrians, but not for other categories of users, such as wheelchair users and cyclists. It was attractive to pedestrians due to adequate provision of paved footpaths although, at some sections, the gradient of the route is steep, and not friendly to users.

With regards to wider network and inter-modal connection, the route appropriately connects Ngong and Langata Roads and bus stages within the route. It also appropriately connects with Ngumo Road, as there is a provision of overpass at the junction of these two roads. This facilitates crossing for those going to Kenyatta Market and Mbagathi District Hospital and residential estates around the area. It further appropriately connects with those crossing the road to residential areas, such as Kibera and Nyayo Highrise, among others. The major change in 2015 is the location of traffic lights at the Ngong Road–Mbagathi Way Junction and the segregation of pedestrians from the motorway, at the same junction which provides further continuity and linkage to Valley Road and Argwings Kodhek. This adds to the quality of the NMT network in this area.

Valley Road

Valley Road is an important connector of the Nairobi CBD to the northern part of the city. It traverses along various institutions, such as the Ethics and Anti-Corruption Commission Offices (EACC) and Christ is the Answer Ministries Church (CITAM), among many others. It is a route also used by Public Service Vehicles (PSVs) originating from Kibera and Kawangware to the CBD. The road is hilly, especially from the EACC offices towards the Ralph Brunche Bridge, and not ideal for cycling. The road does not have provision for cyclists either.

The Valley Road route has a paved footpath of approximately 2.3 km. Other notable NMT facilities along the route are an overpass and pedestrian crossing point. The overpass has an intact surface, with support handrails and is in good condition, at the junction of Valley Road and Ralph Bunche Road. The location of the overpass is strategic, since this is a popular crossing point for people working in this area, and those attending church services, as well as those going to and from the Nairobi Hospital. It also shows a well-paved footpath between Uhuru Highway and Nyerere Road. The footpath is well graded, and the width is around two metres and can accommodate other users, such as those using wheelchairs, as the gradient of the path is nearly flat. A unique feature is the welded and grounded seats near Uhuru Park, where NMT users can sit and relax and have a break, before continuing with their trips.

In terms of coherence, the footpath is not complete. The area immediately after Nyerere Road Junction is not paved and the upper part of Valley Road, around Ralph Bunche Road all the way to Ngong Road, has a discontinuous condition which makes it unattractive and uncomfortable to use. However, the section near Uhuru Highway is pleasant and attractive to NMTs. The connectivity with other inter-modal facilities, such as bus stages, is fairly good, especially around the Ralph Bunche Road area.

Waiyaki Way

Waiyaki Way traverses through Westlands, Kangemi and Uthiru. The distance between Westlands and Uthiru is estimated to be 18 km, with Kangemi to Westlands being 9 km, which is a good cycling distance. Kangemi residents generate a lot of NMT traffic along this route. The route serves another high NMT traffic flow corridor, mostly originating from Kangemi, a low-income settlement area.

Waiyaki Way has a paved footpath measuring around 1.7 km, which is not in good condition. It also has functional flyovers and NMT seating furniture near the Westlands PSV terminus, and an assisted crossing around Agha Khan Primary School. The provision of NMT support facilities is very minimal. This is despite having several offices and shopping centres requiring walking and crossing of the road.

The only notable supportive and functional NMT facility is the overpass at the St. Marks and Kangemi area. Other facilities, such as pedestrian crossings and speed bumps, were not visible along this route, though the one at Safaricom House was clearly marked and visible, hence rating poor in terms of traffic calming measures. However, assisted crossing was available. The footpaths from the Westlands stage are not in the best condition and are very close to the arterial road, which greatly contrasts with ABC Place, which has paved footpaths complete with bollards. Storm drains are also available but not well maintained. At Kangemi Market the footpath is not recognizable due to encroachment by the traders, and the transportation pick-ups forcing pedestrians to deviate to the main highway. This is a route with great potential for NMT infrastructure based on the observed web of NMT movements along the route.

The safety of the majority of the NMT users, especially pedestrians and cyclists, is not guaranteed on this route. Generally, the scenario on this route is prone to conflict between NMT and motorized transport, due to exposure to conflicting situations. It negates most of the principles attributed to design and provision of NMT facilities. The route lacks buffer facilities, thus exposing NMT users to danger except for the overpasses and the traffic lights used as traffic calming measures. Overall, the footpath is not connecting users to destinations and, hence, not attractive to use, especially during the rainy season.

Kiambu Road

Kiambu Road also hosts a lot of institutions, such as the Criminal Investigation Department (CID) headquarters, Kenya Forest Services (KFS) headquarters, and Kiambu Institute of Science and Technology (KIST), which attract both motorized and NMT traffic. Kiambu Road has an unpaved footpath that is 5.6 km long and a flyover. It has an overpass near Muthaiga Golf Club, and there is also good signage and availability of a bus terminus at the same area, hence, no encroachment by public transport vehicles.

The footpath right before Muthaiga Golf Club is well paved and has a marked cycle lane, an easily noticeable addition to what was there in 2006 and 2011. However, the condition changes right after the golf club where the footpath used is dusty and unpaved. This forces pedestrians and cyclists to use road shoulders, which is too close to the arterial road. The footpaths are also heavily disconnected after the CID Headquarters all through Sharks Palace. There are speed bumps at Kenya Forest Institute and at Ridgeways but no signage.

It can, thus, be summarized that the principles of NMT design and provision are widely ignored with regard to the route, except for the one overpass around Muthaiga Golf Club. It is, however, important to note that, due to inadequate provision of the NMT infrastructure coupled with unfavourable geographic terrain, the route is not attractive for cycling. The current width of the road and the number of motorized transport does not encourage the origin to destination type of walking. This requires that the busy points and institutions be well linked with walkways and traffic calming infrastructure. The route has a wide connection to Thika Road and Muthaiga Road on the Nairobi City side and to Kiambu Town, with various bus stages strategically located along the road.

Zone B roads

First Avenue Eastleigh Road

This route connects heavy NMT traffic flow from densely populated areas of Mathare, Eastleigh through Biafra, Bahati towards Jogoo Road and the labour-intensive industrial area (see Figure 7.3).

Figure 7.3 Nairobi NMT provisions on roads in the Eastlands.
Source: Field Survey, 2015.

The terrain is fairly gentle, save for the bridge near Biafra. It has potential for both pedestrians and cyclists. The general condition of the route is fairly good. Approximately 3.2 km of the path is paved and has traffic calming facilities, such as speed bumps, a pedestrian crossing and a footbridge.

The paved footpath goes all the way from Juja Road to First Avenue Eastleigh Roundabout to Jogoo Road Junction, although the entire route is not protected from motorists. Some sections of the footpath are encroached by motorists and informal traders and the cycle lanes have not been spared either. Garbage was also spewed all over footpaths and *matatus* also use the footpath as a terminus, creating total chaos as pedestrians have no option but to use the arterial road. The Garissa Lodge area was the worst hit.

There are speed bumps as calming measures from Eastleigh through Bahati to Jogoo Road Junction with a pedestrian crossing located near Jogoo Road Junction. CCTV cameras and streetlights are also very well distributed from Juja Road Junction all through Eastleigh to Jogoo Road Junction. NMT facilities within the route, such as a footpath and footbridge, are fairly safe and coherent to users, but the level of attractiveness and comfort still rank poorly, since some of the sections are prone to encroachment by motorcars and traders.

Heshima Road

This is the road which traverses various estates, such as Buruburu, Uhuru, Jerusalem, Kimathi and Bahati. It is also a useful route for NMT traffic coming from Kariobangi to the above estates for work as mechanics, dressmakers and other forms of work. Children from the surrounding estates also walk to various schools within this route. Most of the route has a gentle slope, suitable for walking and cycling from origin to destination or as part of inter-modal transfer. Heshima Road has a paved footpath in good condition measuring slightly over half a kilometre, but most sections, approximately 4.2 km, can be classified as fair. In terms of traffic calming measures, the route has provision for speed bumps.

The longest part of the footpath is in fairly good condition. The area which can be termed good is the section between Bahati to Wangu Avenue, where there is a well-paved footpath; it does not have buffers, however, which compromises the safety of NMT users. Several speed bumps are strategically located near Morrison, Kimathi, Dr. Livingston and Uhuru Primary/Secondary Schools. The corridor has several primary schools and the provisions are appropriate for schoolchildren, who largely walk to school.

Incidences of encroachment by traders and *matatus* on the arterial road by NMT are common, due to lack of adequate provision of NMT facilities, especially around Shinyanga Road and First Eastleigh Junctions. Some pedestrians are also using the road shoulders due to inadequate NMT infrastructure in some sections.

The route is fairly served with different types of NMT facilities but they are still inadequate, given that quite a number of people going to the industrial area and the Uhuru and Burma markets from neighbouring low-income areas and informal settlements, such as Kiambiu and Kariobangi, use the route. The design of NMT facilities on this route can be ranked as fairly safe, especially for school-going children and those neighbouring the schools. There is strategic provision of speed bumps, and the route is fairly coherent, although not attractive, since it does not adequately provide for NMT users. There are, however, CCTV cameras at Heshima Road Junction and Jogoo Road.

Mumias Road

The Mumias route is central in networking NMT traffic from Kariobangi, the Outer Ring area and the Buruburu area to several destinations, such as major shopping facilities within the area and on Rabai and Jogoo Roads. It provides a link to Jogoo Road and to the industrial area and other facilities along Jogoo Road. The route is an alternative route to those going to the industrial area from the Kariobangi area, and it is attractive to NMT traffic, including cyclists.

The total length of the footpath, in fair condition, was approximately 1.2 km while more than 4 km was considered poor. Some sections of the footpath around Buruburu in Phase I were paved and slightly spacious compared to some sections around Buruburu Police Station in Phase II. The route has several speed bumps near institutions and the shopping centres to facilitate crossing of the road. On one side of the road, there is no adequate provision of streetlights and zebra crossings. Footpaths along Buruburu Shopping Centre are well paved and complete with bollards. Use of signage has also not been fully maximized, though there was one indicating a speed bump at Buruburu Phase I and at the Church of Jesus Christ area in Buruburu Phase IV.

The route is fairly coherent in terms of completeness of footpath points; it is also fairly safe as some sections are completely separated from the arterial road, e.g. around Buruburu Phase II. The route is fairly attractive and comfortable for pedestrians, but not for cyclists and hand carts. A few errant motorized vehicles, in particular *matatus,* occasionally encroach onto the walkways though. Other challenges include paths obstructed by parked motorcars, narrow sections inhibiting smooth traffic flow of pedestrians and stagnant water which limits the users to the arterial road or pushes them to an unpaved section of the facility.

Rabai Road

The road is fairly short, connecting Jogoo Road with Heshima Road but cutting through Jericho Estate and Market, Metropolitan Hospital, Rabai Road Primary School, Harambee Estate and Uhuru Estate. It is attractive to NMT users, mainly pedestrians, as it is a short distance but connects various useful functions within a short radius. Rabai Road is characterized by a gentle slope, and hence is a suitable route for cycling.

Rabai Road has speed bumps, traffic lights, streetlights and a paved footpath in fair condition measuring approximately 1 km. The footpath is well paved at the ACK Jericho Church, but only on one side, with signage to indicate speed bumps. Metropolitan Hospital area also has a paved footpath with continuous speed bumps. From Jericho Market, the footpath is very dusty and not paved.

Most sections of the route are unpaved; there is only a limited paved path near the National Council of Churches (NCCK) and Jua-kali Industries. The NMT users are occasionally forced to avoid the unpaved path near St. Philips ACK Church, due to poor drainage and a raw sewer, which makes it unsafe to use. A notable new feature in 2015 is CCTV near the Rabai–Jogoo Road Junction. Overall, the road can be said to be unattractive and unsafe to all categories of NMT. It is also incoherent and lacks a smooth continuous network for any NMT.

Jogoo Road

Jogoo Road is a major carriageway connecting the Nairobi CBD and industrial area with various Eastlands neighbourhoods, such as Kaloleni, Makongeni, Makadara, Buruburu and Donholm, among others. It attracts various NMT users, depending on the origin and destination, with some pedestrians and cyclists being those riding from origin to destination, but quite a number of pedestrians using NMT as inter-modal, mostly from the bus terminus to their work stations. Jogoo Road terrain is fairly gentle, and hence suitable for cycling.

Jogoo Road has several NMT facilities, such as a paved footpath measuring approximately 2.8 km, a pedestrian crossing, overpasses and a footbridge. There is a paved footpath from Donholm Roundabout up to Nile Road Junction. There are two overpasses, one near City Stadium and the other one near Uchumi Supermarket. There is also a footbridge near Donholm Roundabout, where Jogoo Road joins Outer Ring Road.

Jogoo Road has several facilities: the most noticeable are the two overpasses which are strategically located near two major markets along this road, namely Uhuru and Burma Markets. Some sections of the road have well-paved footpaths in good condition, complete with bollards. Pedestrian crossings are also very well distributed along the road with marked signage. The storm drainage facilities are poorly maintained, and do not serve their full purpose during the rainy season.

There is encroachment on footpaths along Uhuru Market and City Stadium by traders. Storm drainage facilities have also been affected by makeshift structures put up by traders. The zebra crossing at Rikana Supermarket has completely faded off and it no longer serves its purpose.

Given its characteristic, the Jogoo Road NMT design and principles can be summarized as unsafe, especially for pedestrians, cyclists and hand carts, as there is inadequate provision for these varieties of users. It is not coherent, as exhibited by disjointed paving of the footpaths. It is also unattractive and not comfortable for all types of NMTs. A new notable feature in 2015 is the CCTV cameras distributed strategically along this route, which are expected to assist in recording incidences of accidents involving NMTs.

Lusaka Road

Lusaka Road mainly connects the industrial area to Jogoo Road and Uhuru Highway Junction. It terminates around two major stadiums in Kenya, namely Nairobi City and Nyayo Stadiums. The route connects the industrial area to the major artery leading to residential quarters and major sporting facilities, such as the stadiums just mentioned. It is a busy road for those connecting from Jogoo Road and those walking to Kibera towards the industrial area. It is fairly hilly around the two stadiums but the general terrain is gentle; hence, it is attractive to cycling. There is a paved footpath covering approximately 0.3 km and an unpaved footbridge. The paved footpath section of Lusaka Road is near Jogoo Road Roundabout and a footbridge near City Stadium.

Lusaka Road has very few NMT support facilities with only a small section having a paved footpath. From City Stadium, the footpath is paved and complete with bollards; there is also a terminus on one side, thus eliminating conflict between NMT users and motorists. The area around City Stadium is not paved and motorists block users by parking on pavement along building frontages. This creates conflicts between NMT and motorized traffic.

There is also encroachment by *matatus* on the footpaths. Overall, the design and provision of the NMTs along this route is generally unsafe, incoherent, unattractive and uncomfortable to users.

Conclusion

All 18 corridors had some NMT facilities for pedestrians, attracting a mixture of NMT modes, ranging from pedestrians and cyclists to hand carts, with minimal provision for people with disabilities. Some of the corridors, such as Mbagathi, Kingara and Dennis Pritt, seem to be ahead in provision and management of NMT facilities. Provision of NMT infrastructure along the 18 Nairobi arteries/routes has improved, as compared to the years 2006 and 2011. Some routes have spacious and well-maintained NMT infrastructure, while others are either not well maintained or obstructed by other activities, including garbage dumping. However, the types of infrastructure provided are not uniform and do not fully conform to the NMT design principles of safety, coherence, attractiveness and comfort.

Most of the routes still lack continuous bollards and proper segregation of modes of transport. They strictly follow the arterial road route, making it hard to gauge whether what is planned is the shortest and preferred route for NMT users, or used purely due to their availability. Furthermore, the discontinuous provision of the facilities point at NMT as a mere conduit of other modes of transport, with inter-modal connectivity to bus stages, in relation to origin and destination, which is shown by placements of bollards and zebra crossings near bus stages and markets, and totally absent along the NMT routes. The users' behaviour is also different, with some preferring to use the arterial roads, even in cases where NMT infrastructure is provided elsewhere. This raises the question of suitability of new infrastructure in terms of comfort and connectivity of users to other modes. It further calls for rationalization of allocation of road space across all types of road users, more specifically, NMTs.

Issues of management, such as painting of traffic calming facilities, cleanliness of overpasses and functional streetlights, need to be addressed, together with enforcement, especially with regard to the use of provided facilities and right of way in ensuring that both the NMT and motorized users are adequately protected. This creates a demand for functional NMT infrastructure provision, which observes all the ideal requirements, as stipulated in the draft 2015 Nairobi NMT policy.

While this inventory provides good information, a key principle in NMT design, which is measuring the directness of NMT infrastructure, including assessing the relationship between origin and destination, could not fully be done using mapping and observation methods. However, it was apparent that in some NMT infrastructure routes, it was assumed that NMT was to be used as an access/egress mode and not to be solely used from origin to destination. In the African context, this is not the case and this misconception should be addressed in future NMT planning and implementation.

Notes

1 The 18 routes are the result of Katahira and Engineers International mapping of 2006 in the city of Nairobi.
2 Victoria Transport Policy Institute. 2011. *Pedestrian and bicycle planning: a guide to best practices.*

References

Katahira and Engineers International. 2006. The study on master plan for urban transport in the Nairobi metropolitan area (NUTRANS). Funded by the Japan International Cooperation Agency (JICA).
Republic of Kenya. 2012. Integrated National Transport Policy. Nairobi: Government Printer.
Victoria Transport Policy Institute (VTPI) (2011). Pedestrian and Bicycle Planning: A Guide to Best Practices. Victoria, Canada.

8 An investigation into the effects of NMT facility implementations and upgrades in Cape Town

Jennifer Baufeldt and Marianne Vanderschuren

Introduction and overview

Non-motorized transport (NMT), as a mode of transport, is beneficial and sustainable for both developing (Litman, 2002; Behrens, 2005; Bechstein, 2010; UNEP, 2010; Bogotá Como Vamos, 2014; Verma et al., 2015) and developed countries (Cavill et al., 2008; Dill, 2009; Oja et al., 2011). South African NMT users face various challenges (Behrens and Wilkinson, 2001; Behrens, 2004; City of Cape Town, 2005; NDoT, 2008; Mabunda et al., 2008; Mackenzie et al., 2008; Beukes et al., 2012) that reduce the attractiveness of selecting NMT trips. The main concern is the inadequate provision of NMT facilities, making NMT trips inefficient and dangerous.

One strategy of addressing this challenge is implementing NMT facilities that provide safe, convenient and comfortable routes for NMT users (NDoT, 1998; Pucher et al., 1999; NDoT, 2003; Pucher and Buehler, 2008; Reynolds et al., 2009). By improving the quality of NMT facilities and increasing the number of NMT facilities available to NMT users, the quality of service that NMT users experience will increase (Hook, 2003; City of Johannesburg, 2009; De Waal, 2015) while, at the same time, reducing the levels of concerns regarding safety of NMT trips. These changes should result in fewer NMT fatalities and injuries, as well as an increased number of NMT trips.

In other countries, research (Wright and Montezuma, 2004; Wegman et al., 2007; Pucher, Dill and Handy, 2010) has shown that NMT facilities do, indeed, have a positive impact on NMT users and result in a decrease of fatalities and injuries. However, this is not always the case and there are factors that could impact on the effects that NMT facilities have (Elvik, 2009). In South Africa, there is limited research to illustrate the effects that NMT facilities are having on safety and usage. The research, reported upon in this chapter, was to fill this knowledge gap and determine whether the same established relationships are present in South Africa.

This chapter summarizes this research conducted at the University of Cape Town, which used case study areas to determine the effects of NMT facilities. Where NMT implementations have occurred, the study assessed whether NMT facilities hold the same potential benefits for South African NMT users, as witnessed in developed countries, despite the possible hindering local context.

NMT facilities that have been implemented are investigated in terms of whether they are satisfying the needs of the NMT users. The research explains the, potentially, twofold impact of improving NMT facilities, in terms of safety and public usage of NMT trips.

Firstly, an investigation was conducted into whether improved NMT facilities reduced fatalities and injuries of NMT users. Secondly, a comparison of the volume of NMT trips in the study areas from 2003 to 2013 was conducted, in order to determine whether there had been an increase of NMT trips in the areas that have had upgrades. Additionally, the actual implementations were assessed to determine how successful these implementations were, based on the recommendations presented in the new NMT Facility Guidelines (NDoT, 2015). From these three main investigations, the progress and shortcomings of the NMT implementations were established, regarding the impact these upgrades have had in terms of improving the safety of NMT users, as well as increasing usage of the facilities.

The findings reveal that, to some extent, NMT facilities have had significant impacts on improving the levels of safety of both pedestrians and cyclists in South African settlements. Additionally, NMT facilities do have the potential to improve the quality of service that NMT users experience, based on the qualitative investigations that were conducted. However, while some NMT facility implementations have improved the level of service to NMT users, based on the principles within the NMT Facility Guidelines (NDoT, 2015), much more can be done in terms of improving the design and implementation of NMT facilities, so that the maximum benefit from NMT facility implementations can be generated.

Background

Providing adequate levels of service for NMT users is seen as an important aspect in the literature for other countries (Pucher and Buehler, 2008; Verma et al., 2015), as well as locally (Vanderschuren and Galaria, 2003; NDoT, 2015). In South Africa, the lack of adequate NMT facilities could be seen as hindering the acceptance and support of NMT in South Africa as a sustainable mode of transportation (Gwala, 2007; Ribbens et al., 2008; City of Cape Town, 2009). The current prioritization of NMT facilities seems to be much lower than what is necessary to make substantial progress, in terms of providing adequate implementation of facilities, to ensure safe, convenient and comfortable NMT trips in South Africa (Vanderschuren and Galaria, 2003; Behrens, 2004; NDoT, 2015).

There are various potential reasons why there are inadequate facilities. These range from insufficiently trained stakeholders and practitioners, to more basic issues, such as insufficient funding frameworks, that allow for the appropriate NMT facilities to be built and maintained (City of Cape Town, 2009).

These, and other issues regarding NMT, are being addressed by various local governments and municipalities through measures such as master plans and guidelines (Visser et al., 2003; City of Cape Town, 2005; Cape Winelands District Municipality, 2009; City of Johannesburg, 2009). However, an integrated NMT approach, which takes into account the various stakeholders in both the public and private environments, is still lacking. The lack of an integrated approach can be seen in the many new developments, upgrades of current road facilities and new Public Transport (PT) facilities and services not taking into account the needs of NMT users. The NDoT hopes that the newly drafted NMT Facility Guidelines will help address this issue, by encouraging practitioners and stakeholders to include NMT facilities through providing comprehensive and practical guidelines (NDoT, 2015). However, as the guidelines are not legally required, the effectiveness of this approach is questionable and relies heavily on how the guidelines are distributed and presented to the relevant stakeholders and general public. Additionally, the relevant stakeholders and practitioners that are responsible for designing and implementing NMT facilities will need to be motivated to adopt the new NMT Facility Guidelines into their work.

The consequences of inadequate provision of NMT facilities include illegal and dangerous travel behaviour (both by NMT users and motorists), as well as increased dependency on motorized transport trips. By not providing adequate facilities for NMT trips, individuals switch to motorized forms of transport as soon as these modes are available or affordable for them. This switch to motorized transport is problematic as it results in increased levels of congestion (Gwilliam, 2003; Walters, 2008) and the various negative externalities associated with private motorized transportation (Massink et al., 2011; WHO, 2011; Peden et al., 2013), as well as resulting in PT becoming less viable, due to a declining number of passengers. Additionally, motorized transport use reduces the amount of physical activity of individuals on a daily basis (Elvik, 2000; De Hartog et al., 2010; Gomez et al., 2015), which further worsens the levels of health of South Africans (Mayosi et al., 2009; Health Systems Trust, 2015; Vanderschuren et al., 2015) which, in turn, increases the cost of health care for South Africa (City of Cape Town, 2014).

An additional consequence of inadequate NMT facilities is the high level of inequity that is reflected in the space allocation to the different road users (Litman, 2002; Van Wee, 2011; Lucas, 2011). Individuals with higher incomes are dominating public road spaces in South Africa with private cars (Behrens, 2004; Van Wee, 2011) and vulnerable road users, such as pedestrians and cyclists, are being poorly provided for (City of Cape Town, 2005, 2009). Addressing inequality throughout South Africa, especially in public spaces, is an important aspect of transformation in South Africa (Özler, 2007). In the Constitution of South Africa, as well as the Bill of Rights, improving levels of equality regardless of income, race, age or gender is a central theme. Therefore, addressing the needs of NMT users is an important aspect of upholding these rights in a practical manner, which could have significant positive impacts for society as a whole (NDoT, 2008; City of Cape Town, 2009).

National guidelines, regarding NMT facilities in South Africa, were first presented under the title of *Pedestrian and Bicycle Facility Guidelines* (NDoT, 2003). However, as there was no legal requirement for the guidelines to be followed, the degree to which the *Pedestrian and Bicycle Facility Guidelines* (2003) were adopted and implemented is unclear, especially on a national level. Based on the progress of NMT implementations, the success of the guidelines is seen to be limited. However, the *Pedestrian and Bicycle Facility Guidelines* (2003) provided a well-researched and comprehensive foundation, which was used as a starting point for the recently revised and expanded NMT Facility Guidelines (NDoT, 2015).

The revision and expansion of the guidelines included new concepts and best practices, such as Universal Access, as well as addressing other identified gaps (NDoT, 2015). The updating of the guidelines also aimed to make the facility guidelines more user-friendly and more applicable to the challenges that face South Africans today.

By improving and expanding the NMT Facility Guidelines (NDoT, 2015), the National Department of Transport hopes that more stakeholders and practitioners will consult and, therefore, implement facilities that adhere to and reflect the principles and designs of the guidelines, thus improving the consistency and quality of NMT facilities implemented in South Africa. Whether the greater awareness and advocacy of NMT users in South Africa will be sufficient to increase the compliance of practitioners to the new NMT Facility Guidelines will need to be investigated in the future in order to determine the effectiveness of this non-legal approach to the guidelines. This has already been raised as a concern by many experts and stakeholders, both in government, as well as in practice.

The implementation of NMT facilities has been identified to be a major weak point of NMT in South Africa. This was found to be consistently mentioned in the literature reviewed, as well as through stakeholder engagements during the revision of the NMT Facility Guidelines. The NMT network, throughout the country, is very limited and often does not incorporate adequate designs and principles (Vanderschuren and Galaria, 2003; Behrens, 2005; Ribbens et al., 2008; NDoT, 2008; City of Cape Town, 2009). The lack of NMT facilities could further be seen as limiting the effectiveness of other NMT initiatives, especially those that are aimed at encouraging individuals switching to NMT modes (Behrens, 2005; Pucher et al., 2010; Heinen et al., 2010; Verma et al., 2015).

The focus of this work has been to investigate NMT facilities in South Africa to explore the effects that implementing adequate NMT facilities will have on NMT and NMT users. Therefore, the aim of this research was to identify the specific effects that existing NMT facilities have had, as well as exploring possible issues with the manner in which NMT facilities have been rolled out. This has been investigated in detail with Cape Town serving as the selected case study (for further details of the policy and overarching bicycle-specific context in Cape Town, see Chapter 13).

Method

The focus of the research was the safety and usage of NMT in relation to the implementation of adequate NMT facilities in South Africa. Due to the nature of the research questions, a mixed-methods approach, featuring both qualitative and quantitative data, was used. This approach allowed a comprehensive and multi-perspective investigation of NMT implementations to be presented (Olsen, 2004; Yeasmin and Rahman, 2012).

One of the motivating factors behind this research was taking into account the local context in terms of the appropriateness of NMT facilities. Additionally, the rollouts of NMT facilities have occurred at various times and in different areas, which would have influenced the impacts. Lastly, there are aspects that are of interest, which are difficult to measure in quantitative terms, namely how well the NMT facilities that have been implemented are serving the needs of the NMT users. The research aimed to investigate whether the NMT facilities that have been implemented are of a good quality, which would improve the safety of NMT users and encourage NMT trips. Due to the importance of context and qualitative features of the study, a *case study design* was considered an appropriate option (Greene et al., 1989; Baxter and Jack, 2008; Andrade, 2009).

The main aim of these investigations is to determine if there is a link between the NMT implementations and increases in the number of NMT users. This was then compared with the findings, regarding safety and the quality of implementation, to see if volumes of NMT trips could have been affected by a high incident of NMT fatalities and injuries, or a poor quality of NMT facility implementations.

The third element of the research focused on determining the quality of the NMT facilities. This relied mostly on visual aids, namely Google Maps images, both current and historic, to evaluate the success of these facilities. The NMT facilities were evaluated against the criteria and recommendations as set out in the recently published National NMT Facility Guidelines (NDoT, 2015). The third investigation also used aspects of both the safety investigations and NMT trips investigations, to refine the last type of investigation.

For the safety aspects, the trends of NMT fatalities and injuries were used as the criteria to determine what types of impact the implementation of NMT facilities were having. The second element, which is concerned with the usage or number of NMT trips, compared the 2003 and 2013 National Household Travel Surveys (STATSSA, 2005, 2014). The case study areas and control areas were compared in terms of increases in volumes of NMT trips. The criteria for this element was the volume of NMT trips. Finally, for the last element, which was the quantitative evaluation of the NMT facilities, a scoring system was used where the NMT facilities were compared against the recommendations in the NMT Facility Guidelines (NDoT, 2015) and given a rating.

Through the use of several case study areas of NMT implementations in Cape Town, the degree to which these implementations could encourage NMT was assessed by comparing what had been implemented to what had been recommended in the NMT Facility Guidelines.

A selection of NMT facility implementations was identified and grouped according to location or suburb, so that nearby NMT facilities created a single case study area. From these groupings, the areas that had the most concentrated NMT facility implementations were then selected as case study areas. The groupings and lengths of roads that were upgraded are shown in Figure 8.1. Areas with limited NMT facilities were excluded, as it was assumed that the impacts would be too small to be reflected in the data.

Control areas were then selected out of the other suburbs of Cape Town, which had no NMT implementations during the period of time that would be investigated. From these data sets (Transport for Cape Town, 2015), graphs were drawn to illustrate the trends of fatalities and injuries in the form of Equivalent Accident Numbers (EANs) for both the case study areas and the control areas (see Figure 8.2). Pedestrians and cyclists were investigated separately, as these users have different facility needs. Therefore, the facilities that have been implemented may only be appropriate for one type of NMT user, and not for the other type of NMT user.

After the graphs had been drawn, statistical analysis of the changes in the EANs, within each case study and control area, were done. Additionally, whether the case study areas and the control areas were significantly different, overall, was also investigated. This was determined by conducting T-tests on the case study areas and the control areas for pedestrian EANs and cyclist EANs. Tables 8.1 and 8.2 show the summary of the T-tests that were conducted for the case study areas and the control areas, respectively.

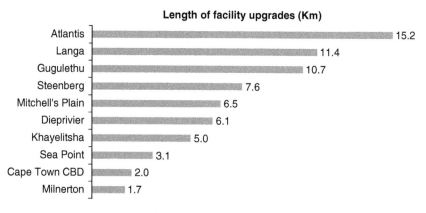

Figure 8.1 Case study areas with non-motorized transport facility upgrade lengths (Km).
Source: Baufeldt, 2016.

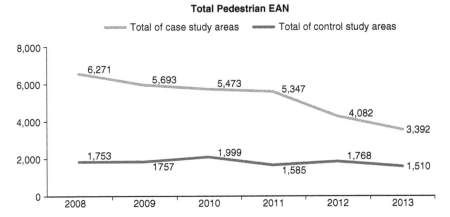

Figure 8.2 Overall trends of pedestrian EAN in case study areas vs. control study
areas.

Source: Baufeldt, 2016.

As seen in the T-tests, the NMT facilities were, generally, more significant for pedestrians than for cyclists. The newly drafted NMT Facility Guidelines provided clear outlines of what would be necessary for facilities to be viewed as adequate and attractive for NMT users. Therefore, using these guidelines, a qualitative rubric was drafted, which focused on key principles of the NMT Facility Guidelines (NDoT, 2015).

Based on the findings of the first quantitative investigation, regarding EANs of NMT users, the best and worst performing case study areas and control areas were selected. Within these areas, routes were selected and Google Maps' Street View was then used to scan through the selected routes to determine what rating would be applicable per principle of the rubric per route. From these numeric ratings, principles that are currently poorly implemented or currently successfully implemented could be determined. This helps to guide practitioners and local governments on which principles need more urgent attention and how established principles could be improved further.

An example of one of the better quality pedestrian implementations was Spine Road, in Khayelitsha. This was illustrated in the images captured from Google Maps, as shown in Figure 8.3.

As highlighted by the circles in Figure 8.3, the upgrades have provided adequate space for pedestrians and have improved the quality of the environment through landscaping and appropriate street furniture. However, as shown by the circle in Figure 8.3, there is a lack of allocated cycling space or separation between cyclists and motorized traffic. This speaks to the EAN T-tests of this area, where the improvement for pedestrians was significant, while the safety improvement for the cyclists in this area was not. By visually assessing the NMT implementations, some of the key failures

Table 8.1 Summary of T-test results per case study areas

Area	Date of upgrade	Pedestrian			Cyclist		
		P-value	T-critical	Significant	P-value	T-critical	Significant
1 Atlantis	2012/13	0.0011	4.540	Yes	0.0093	6.96	Yes
2 Langa	2011/13	0.0404	6.964	No	0.0352	31.82	No
3 Gugulethu	2008/09 vs. 2012/13	0.0564	31.820	No	0.1211	31.82	No
4 Steenberg	2012/14	0.000163	4.5407	Yes	0.0015	6.96	Yes
5 Mitchell's Plain	2012/13	0.0001	4.5407	Yes	0.0031	6.96	Yes
6 Dieprivier	2012/13	0.0004	4.5407	Yes	0.0022	6.96	Yes
7 Khayelitsha	2008/09 vs. 2012/13	0.0086	31.8205	Yes	0.0352	31.82	No
8 Sea Point	2008/09 vs. 2012/13	0.0067	31.8205	Yes	0.0583	31.82	No
9 Cape Town CBD	2010/12	0.0176	31.8205	No	0.0127	31.82	No
10 Milnerton	2011/12	0.000263	4.4507	Yes	0.0110	31.82	No

Source: Baufeldt, 2016.

Table 8.2 Summary of T-test results per control area

Area	Date of upgrade	Pedestrian			Cyclist		
		P-value	T-critical	Significant	P-value	T-critical	Significant
Control 1: Philippi	2008/09 vs. 2012/13	0.0196	31.8205	No	0.0605	31.8	No
Control 2: Bishop Lavis	2008/09 vs. 2012/13	0.0252	31.8205	No	0.0736	31.82	No
Control 3: Ravensmead	2008/09 vs. 2012/13	0.0094	31.8205	Yes	0.1169	31.82	No
Control 4: Kirstenhof	2008/09 vs. 2012/13	0.0519	31.8205	No	0.0828	31.82	No
Control 5: Camps Bay	2008/09 vs. 2012/13	0.0017	6.9646	Yes	0.0672	31.82	No
Control 6: Bothasig	2008/09 vs. 2012/13	0.0212	31.8205	No	0.0946	31.82	No

Source: Baufeldt, 2016.

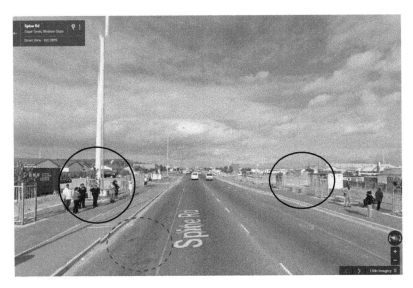

Figure 8.3 Spine Road infrastructure assessment.
Source: Based on Google Street View, by Baufeldt, 2016.

and successes of the implementations can be seen relatively easily. This is one way that practitioners could use past NMT facility upgrades to improve the quality of NMT facilities, by first critiquing existing implementations, before repeating similar errors in design or implementation.

Findings

Fragmented NMT improvements

In Cape Town, the locations of NMT facilities are spread out through the Cape Town area, with many of the NMT facility implementations being isolated from other NMT facility implementations. While this strategy has meant that more areas received improvements, it has also meant that the NMT network is likely to remain fragmented for a longer period of time before the different NMT implementations begin to link together. Considering that NMT users are likely to be deterred by one section of a trip that is problematic, more focus should be placed on improving the connectivity of the NMT network (Pretorius, 2015) than the range of the NMT network. This is especially something to consider when there is a limited budget for the development of NMT facilities in South Africa.

The nature of what was implemented depended slightly on the local context. However, generally, the NMT facility implementations aimed to improve pedestrian and cycling paths, as well as the urban environment, by improving landscaping features and the quality of facilities at intersections and applying principles of Universal Access to the designs thereof.

Significant room for improvement

The implemented NMT facilities in the case study areas showed significant impact at reducing the levels of Equivalent Accident Numbers (EANs) of NMT users. Thereby, the NMT facility implementations proved to successfully improve the levels of safety. However, while there have been significant improvements in the case study areas, in terms of improving safety, when compared to the control areas, the NMT facility implementations still have significant room for improvement.

The majority of NMT facilities, which have been implemented in the case study areas, cater well for pedestrians, but the needs of cyclists are not fully understood or adequately provided for. This was, also, found to be the case in terms of the impact that the NMT facilities had on the number of NMT trips. While for pedestrians the investigations indicated that the case study TAZs (Transport Analysis Zones) had noticeably more pedestrian trips than the control TAZs, for cyclists there was no noticeable impact on the number of cycling trips. This indicated that the cycling facilities had had limited or no impact in terms of encouraging the number of trips.

By addressing the gaps in the NMT facilities and improving the designs and implementations of NMT facilities, NMT will not only become a safer manner of commuting, but a more attractive and efficient manner of commuting too.

Insufficient attention to unique NMT needs

Considering the variation of NMT facility implementations seen in both the case study areas, as well as the control areas, there is not one answer for this question. Some case study implementations, such as Khayelitsha, did have evidence that the local context had been taken into account by physically completely separating the NMT routes (pedestrian and cyclists) from motorized traffic. However, in other case study areas, such as Sea Point, the lack of clear allocated cycling routes has resulted in many cyclists still using the motorized road space. Considering that there are many recreational cyclists in this area, the needs of cyclists should have been made more of a priority than what can be seen in the designs. The poor qualitative results of Sea Point align with what was seen in the quantitative investigation of Sea Point. Despite having very recent NMT facility implementations, the EANs for NMT users (particularly cyclists) have actually increased over the recent years.

On a national level, the number of NMT trips is generally declining. The only exception to this is the case studies (TAZs) for pedestrian trips, which have shown positive increases in the number of trips. This is shown in Figure 8.4, where the case study areas are shown in dark blue and the control areas are in green.

This is indicative that local challenges have not been appropriately addressed within the design of the NMT facilities, while also showing that, if appropriate facilities are installed, as per the requirements of the users, positive change can be made.

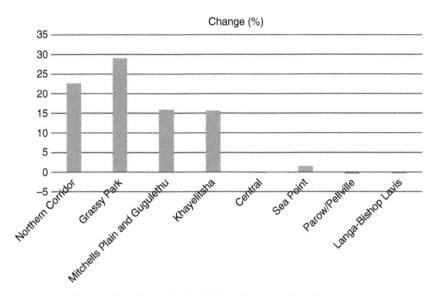

Figure 8.4 Change of walking trips in TAZs of case study and control areas.
Source: Baufeldt, 2016.

Therefore, more can be done in terms of improving NMT facility implementations in South Africa, so that local challenges and needs of NMT users
could be better accommodated, in a more consistent manner. By adopting
the NMT Facility Guidelines (NDoT, 2015), it is more likely that this will be
done in a more successful manner.

In conclusion, while there is still much more to be done, there has been
improvement in NMT facilities over the years. The focus, going forward,
should be to increase the consistency and quality of the NMT facilities
that are maintained and improved upon, as well as increasing the reach
and connectivity of the NMT network, for both pedestrians and cyclists,
accordingly.

References

Andrade, A.D. 2009. Interpretive research aiming at theory building: adopting and
adapting the case study design. *The Qualitative Report*, 14(1): 42–60. Retrieved
from http://nsuworks.nova.edu/tqr/vol14/iss1/3.

Baufeldt, J.L. 2016. *Investigation into the effects of non-motorized transport facility
implementations and upgrades in urban South Africa.* Msc Thesis, University of
Cape Town, Available at: http://hdl.handle.net/11427/20467.

Baxter, P. and Jack, S. 2008. Qualitative case study methodology: study design and
implementation for novice researchers. *The Qualitative Report*, 13(4): 544–559,
Available at: www.nova.edu/ssss/QR/QR13-4/baxter.pdf.

Bechstein, E. 2010. Cycling as a supplementary mode to public transport: a case study
of low income commuters in South Africa. In 29th Southern African Transport
Conference (SATC, August 2010), pp. 33–41. Pretoria, South Africa.

Behrens, R. and Wilkinson, P. 2001. *South African urban passenger transport policy and planning practice, with specific reference to metropolitan Cape Town.* Urban Transport Research Group Faculty of Engineering and the Built Environment University of Cape, Working Paper No. 4, July 2001, South Africa.

Behrens, R. 2004. Understanding travel needs of the poor: towards improved travel analysis practices in South Africa. *Transport Reviews: A Transnational Transdisciplinary Journal*, 24(3), 317–336. DOI: 10.1080/0144164032000138779.

Behrens, R. 2005. Accommodating walking as a travel mode in South African cities: towards improved neighbourhood movement network design practices. *Planning, Practice and Research*, 20(2): 163–182.

Beukes, E., et al. 2012. Using an area-wide analysis of contextual data to prioritize NMT infrastructure projects: case-study Cape Town. *Conference CODATU XV: The role of urban mobility in (re)shaping cities*, 22–25 October 2012, Addis Ababa, Ethiopia.

Bogotá Como Vamos. 2014. *Resultados de la Encuesta de Percepción Bogotá Cómo Vamos 2014*. Bogotá.

Cape Winelands District Municipality. 2009. *Non-motorized transport framework plan towards master planning: version 1.1.* South Africa.

Cavill, N., Kahlmeier, S., Rutter, H., Racioppi, F., & Oja, P. 2008. Economic analyses of transport infrastructure and policies including health effects related to cycling and walking: a systematic review. *Transport Policy*, 15(5): 291–304.

City of Cape Town. 2005. *NMT policy and strategy, volume 1: status quo assessment*, October 2005, Cape Town, South Africa.

City of Cape Town. 2009. *Integrated transport plan for the City of Cape Town: 2006 to 2011*. Review and update October 2009. City of Cape Town.

City of Cape Town. 2014. *State of Cape Town report: summaries*. Cape Town, South Africa.

City of Johannesburg. 2009. *Framework for non-motorised transport.* City of Johannesburg, South Africa.

De Hartog, J.J., Boogaard, H., Nijland, H., and Hoek, G. 2010. Do the health benefits of cycling outweigh the risks?, *Environmental Health Perspectives*, 1109–1116.

De Waal, L. 2015. Interview with Louis de Waal at BEN Bikes. Cape Town, South Africa.

Dill, J. 2009. Bicycling for transportation and health: the role of infrastructure. *Journal of Public Health Policy*, S95–S110.

Elvik, R. 2000. Which are the relevant costs and benefits of road safety measures designed for pedestrians and cyclists. *Accident Analysis & Prevention*, 32(1): 37–45.

Elvik, R. 2009. The non-linearity of risk and the promotion of environmentally sustainable transport. *Accident Analysis and Prevention*. 41: 849–855.

Gomez, L. F., et al., 2015. Urban environment interventions linked to the promotion of physical activity: A mixed methods study applied to the urban context of Latin America. *Social Science & Medicine*, 131: 18–30.

Greene, J.C., Caracelli, V.J., & Graham, W.F. 1989. Toward a conceptual framework for mixed-method evaluation designs. *Educational Evaluation and Policy Analysis*, 11(3): 255–274.

Gwala, S. 2007. Urban non-motorised transport (NMT): a critical look at the development of urban NMT policy and planning mechanisms in South Africa from 1996–2006. *Proceedings of the 26th Southern African Transport Conference.* ISBN Number: 1-920-01702-X, Document Transformation Technologies cc, 9–12 July 2007, Pretoria, South Africa.

Gwilliam, K. 2003. Urban transport in developing countries. *Transport Reviews*, 23(2): 197–216.

Health Systems Trust 2015. *More South African adults now die from obesity than from poverty*. Retrieved from: www.hst.org.za/news/more-south-african-adults-now-die-obesity-poverty (accessed 13 November 2015).

Heinen, E., van Wee, B. and Maat, K. 2010. Commuting by bicycle: an overview of the literature. *Transport Reviews*, 30(1): 59–96.

Hook, W. 2003. Preserving and expanding the role of non-motorised transport. In Fjellstrom, K. (ed), *Sustainable transport: A sourcebook for policy-makers in developing cities*. Deutsche Gesellschaft für Technische Zusammenarbeit (GTZ) (Section 1: Benefits of a greater role for non-motorised transport).

Litman, T. 2002. Evaluating transportation equity. *World Transport Policy and Practice*, 8(2): 50–65.

Lucas, K. 2011. Making the connections between transport disadvantage and the social exclusion of low income populations in the Tshwane region of South Africa. *Journal of Transport Geography*, 19(6): 1320–1334.

Mabunda, M.M. et al. 2008. Magnitude and categories of pedestrian fatalities in South Africa. *Accident Analysis and Prevention*, 40: 586–593.

MacKenzie, S., Seedat, M., Swart, L. and Mabunda, M. 2008. *Pedestrian injury in South Africa: focusing intervention efforts on priority pedestrian groups and hazardous places*. Crime, Violence and Injury Prevention in South Africa. Tygerberg, SA: Medical Research Council-University of South Africa.

Massink, R., Zuidgeest, M., Rijnsburger, J., Sarmiento, O. L. and Van Maarseveen, M. (2011, May). The climate value of cycling. *Natural Resources Forum*, 35(2): 100–111.

Mayosi, B. M., Flisher, A. J., Lalloo, U. G., Sitas, F., Tollman, S. M. and Bradshaw, D. 2009. The burden of non-communicable diseases in South Africa. *The Lancet*, 374(9693): 934–947.

National Department of Transport (NDoT). 1998. *The moving South Africa transport strategy*. South Africa: Author.

National Department of Transport (NDoT). 2003. *Draft of pedestrian and bicycle Facility Guidelines*. South Africa: Author.

National Department of Transport (NDoT). 2008. *Draft of non-motorised transport policy*. South Africa: Author.

National Department of Transport (NDoT). 2015. *Final draft version of NMT Facility Guidelines*. South Africa: Author.

Oja, P., Titze, S., Bauman, A., De Geus, B., Krenn, P., Reger-Nash, B. and Kohlberger, T. 2011. Health benefits of cycling: a systematic review. *Scandinavian Journal of Medicine & Science in Sports*, 21(4): 496–509.

Olsen, W. 2004. Triangulation in social research: qualitative and quantitative methods can really be mixed. *Developments in Sociology*, 20: 103–118.

Özler, B. 2007. Not separate, not equal: poverty and inequality in post-apartheid South Africa. *Economic Development and Cultural Change*, 55(3): 487–529.

Peden, M., Kobusingye, O. and Monono, M.E. 2013. Africa's roads - the deadliest in the world. *SAMJ: South African Medical Journal*, 103(4): 228–229.

Pretorius, L. 2015. Telephonic interview regarding NMT in Cape Town.

Pucher, J. and Buehler, R. 2008. Making cycling irresistible: lessons from the Netherlands, Denmark and Germany. *Transport Reviews: A Transnational Transdisciplinary Journal*, 28(4): 495–528, DOI: 10.1080/01441640701806612.

Pucher, J., Dill, J. and Handy, S. 2010. Infrastructure, programs, and policies to increase bicycling: an international review. *Preventive Medicine*, 50: S106–S125.

Pucher, J., Komanoff, C. and Schimek, P. 1999. Bicycling renaissance in North America? Recent trends and alternative policies to promote bicycling. *Transportation Research, Part A*, 33: 625–654.

Reynolds, C.C., Harris, M.A., Teschke, K., Cripton, P.A. and Winters, M. 2009. The impact of transportation infrastructure on bicycling injuries and crashes: a review of the literature. *Environmental Health*, 8(1): 47.

Ribbens, H., et al. 2008. Impact of an adequate road environment on the safety of non-motorised road users. *Crime, violence and injury prevention in South Africa: Data to action*, pp. 48–69. Tygerberg, South Africa: Medical Research Council, University of South Africa.

Statistics South Africa (STATSSA). 2005. *The First South African National Household Travel Survey 2003.* Technical report, Pretoria, South Africa.

Statistics South Africa (STATSSA). 2014. *National Household Travel Survey: February to March 2013.* Statistical release P0320, Pretoria, South Africa.

Transport for Cape Town (TCT). 2015. Transport for Cape Town: Home. Retrieved from: www.tct.gov.za (accessed 24 November 2015).

United Nations Environment Programme (UNEP). 2010. *Share the road: investment in walking and cycling road infrastructure.* UNEP, November.

Vanderschuren, M.J., and Galaria, S. 2003. Can the post-apartheid South African city move towards accessibility, equity and sustainability? *International Social Science Journal,* 55(176): 265–277.

Vanderschuren, M.J., Baufeldt, J.L. and Phayane, S. 2015. Mobility barriers for older persons and people with Universal Design needs in South Africa. 14th International Conference on Mobility and Transport for Elderly and Disabled Persons, TRANSED, Lisbon, Portugal.

Van Wee, B. 2011. *Transport and ethics: ethics and the evaluation of transport policies and projects.* Delft University of Technology, The Netherlands. Edward Elgar Publishing Limited, ISBN 978184980641.

Verma, P.D., Valderrama, J.S.L., and Pardo, C. 2015. *Bogotá 2014 bicycle account,* Despacio. ISBN: 978-958-57674-6-1.

Visser, D. et al. 2003. Gautrans guidelines for the provision of pedestrian and bicycle facilities on provincial roads in Gauteng. Proceedings of the 22nd Southern African Transport Conference, 14–16 July 2003. ISBN Number: 0-958-46096-5. Pretoria, South Africa.

Walters, J. 2008. Overview of public transport policy developments in South Africa. *Research in Transportation Economics,* 22(1): 98–108.

Wegman, F., Aarts, L. and Bax, C. 2007. Advancing sustainable safety: national road safety outlook for The Netherlands for 2005–2020. *Safety Science,* 46: 323–343.

World Health Organisation (WHO). 2011. *Decade of action for road safety 2011–2020.* Global Launch (brochure), WHO/NMH/VIP11.08. Available at: www.who.int/roadsafety/publications/global_launch.pdf?ua=1.

Wright, L. and Montezuma, R. 2004. Reclaiming public space: the economic, environmental, and social impacts of Bogota's transformation. Cities for People Conference Walk21, June, Copenhagen, Denmark.

Yeasmin, S. and Rahman, K. F. 2012. Triangulation research method as the tool of social science research. *Bup Journal,* 1(1): 154–163.

9 Access and mobility

Multi-modal approaches to transport infrastructure planning

*Edward Beukes, Marianne Vanderschuren and
Mark Zuidgeest*

Introduction

At the heart of transport planning, as it is practiced today, lies the idea that
motorized traffic is best kept separate from other modes of transport, and
that its interface with people can be controlled by a hierarchy of roads defined
by a road classification system (Forbes, 1999; O'Flaherty, 1997). The origins
of the traditional functional classification system, developed in the UK and
the US, can be seen in the *Radburn Layout*, developed in the US in the late
1920s, for the unincorporated new town of Radburn in Fair Lawn, New
Jersey, by architects Clarence Stein and Henry Wright (Lillebye, 1996). One of
the major features of the plan was a hierarchical road network, developed to
separate pedestrians and motor vehicles. All fast-moving traffic was restricted
to feeder roads, and the houses were clustered on cul-de-sacs, with one side
accessible from the street, while the other side opened onto communal gardens that had pathways leading to a central park.

The concept was formalized in 1963, when the report *Traffic in towns*
was published (Buchanan, 1963). This report contained a comprehensive
vision for the design of towns and villages in a highly motorized society.
A distinction was made between roads having a traffic flow function, and
roads that give access to destinations. Buchanan argued that, within access
roads, traffic should be of minor importance relative to the environment
(which relates to his concept of the 'environmental area') and, in every
area, at least the maximum acceptable traffic capacity had to be determined
(see Figure 9.1).

Buchanan's ideas have been widely adopted, and play a critical role in
transport planning today. Marshall (2004) notes that road hierarchy has been
an influential and, often, dominating factor in determining the character of
modern urban layouts. In *Streets and patterns*, Marshall (2005) argues, however, that road hierarchy, as developed from Buchanan's ideas, has often been
criticized for resulting in dull or dysfunctional road-dominated layouts, lacking in 'urbanity or sense of place', but he concedes that this has more to do
with the inappropriate application of the concepts behind road hierarchies,
than with Buchanan's vision of the environmental area.

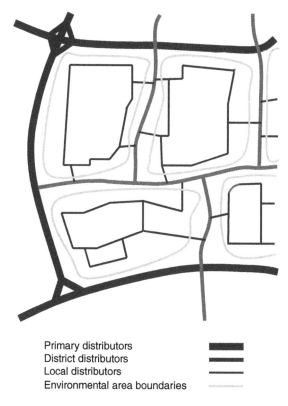

Primary distributors   ▬▬▬
District distributors   ▬▬▬
Local distributors   ▬▬▬
Environmental area boundaries

Figure 9.1 Functional categorization of roads according to Buchanan
Source: Buchanan (1963).

Over the years Buchanan's original ideas have been elaborated upon to suit the needs of the institutions that adopted them. This is illustrated by the approach to road hierarchy as described in *A Policy on the Geometric Design of Highways and Streets* [American Association of State Highway and Transportation Officials (AASHTO), 2004]. This document describes the relationship between 'mobility' and 'access' as an inversely proportional sinusoidal curve. In this conceptualization, all streets fall somewhere along a continuum between mobility-only streets and access-only streets. This view defines the role that the street plays in the network, and forms the theoretical basis for the hierarchy developed, which is now used extensively all over the world. In South Africa, as a case in point for Africa, the interpretation of these ideas can be seen in the still widely used Urban Transport Guidelines Series of road design manuals [Committee of Land Transport Officials (COLTO), 1986]. Figure 9.2 shows how Buchanan's ideas, as described in the AASHTO document, have been translated into a road hierarchy for practical application in road design in South Africa.

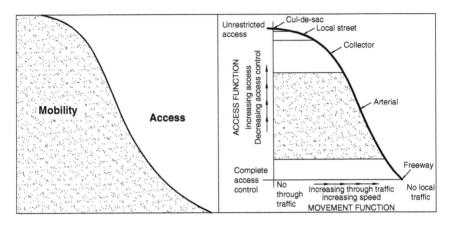

Figure 9.2 The relationship between 'mobility' and 'access' according to AASHTO
            (left) and the translation into road hierarchy in South Africa (right).

Sources: American Association of State Highway and Transportation Officials (AASHTO)
(2004) and Committee of Land Transport Officials (COLTO) (1986).

Transport planning and the provision of infrastructure

The extrapolation of Buchanan's view of road hierarchy into a continuum
of function is problematic, as streets that do not fall neatly along this curve
cannot be accommodated by the classification system. Consequently, there
is no place for streets with high mobility, as well as access functions (many
urban arterial roads fall into this category), or low mobility and access functions (many peri-urban streets fall into this category). As a result, although
the function and, consequently, the infrastructure that is required to serve
that function, is clear towards the extremities of the curve, many streets fall
somewhere towards the centre of the curve, where the precise implications of
the functional relationship is harder to pin down.

In South Africa, and the developing world in general, the approach presents additional difficulties, since roads in these countries are often required
to accommodate a range of functions outside of the traditional scope of
considerations. Urban streets may be required to perform a variety of civic,
ceremonial, political, cultural and social roles, as well as commercial and economic roles, in addition to their movement and access roles (Svensson, 2004).
This multiplicity of roles implies that the functions performed by the road,
and the needs of those who are expected to use it, may, in developing countries, be significantly more complex than what is traditionally understood to
be the case.

The problems brought about by inappropriate transport planning are
amplified in the developing world. According to Iaych et al. (2009), South
Africa has the third highest number of road deaths per 100,000 population
in the world. In South Africa, pedestrians make up the bulk of road accident

fatalities (Mabunda et al., 2008). These figures are cause for great concern amongst decision makers and are the subject of much debate in the public sphere. The National Road Safety Strategy (NDoT, 2006) report proposes a number of strategies to address these problems, which fall into the following categories: a general improvement of law enforcement measures; enhanced road user education campaigns; and the expanded implementation of traffic calming schemes.

In terms of the engineering and road design opportunities for improving road safety, the document has the following to say (see NDoT, 2006, p. 35):

> *There is not a single site in South Africa where more than 1% of crashes occur. Therefore, even if that site is remediated by engineering methods, only a maximum of 1% of crash reduction will occur. Putting effort into behaviour and attitude change is, therefore, more beneficial. Identification of hazardous locations (stretches of road) are, however, still a priority, so that enforcement activities can be concentrated on those areas, during the most dangerous times of the day, and engineering solutions can be explored.*

Considering that more than 18,000 people are killed on South African roads each year (Msemburi et al., 2014), if there were a single location that accounted for 1% of all of these fatalities, let alone all fatal and non-fatal crashes, this would surely be a major cause for concern and, possibly, legal inquiry relating to professional negligence. Surely the benchmark for considering the possibility of a design flaw cannot be that a location must account for more than 1% of all crashes, or even 0.1% of all crashes.

These considerations notwithstanding, the statement also dismisses the potential of road design to effect a change in road user behaviour, and to proactively limit the number of fatalities that occur. The position implied is that there is very little wrong with the way roads are planned or designed. The problem, instead, lies with the way roads are used (or misused).

This view, however, is in conflict with the reality of both trip making and driver behaviour in South Africa. Walking, as both a main and a secondary mode, is very important in South Africa, and a large proportion of road users walk long distances to reach workplaces, public service centres and public transport stops (see NDoT, 2005).

Behrens (2005) notes that both mean and 95th percentile walking trip lengths, in South African cities, are considerably longer than the maximum walking trip length conventionally assumed in practice (+800 m). He holds that this contradicts assumptions regarding the localized nature of walking trips within local neighbourhoods (as initially proposed by Buchanan and assumed in the traditional road hierarchy philosophies), because a significant proportion of observed South African walking trip lengths exceed conventional parallel arterial or district distributor frequencies (of 1,500 and 2,000 m).

Mode choice in South Africa is often dictated by income. Specifically, lower income people are, generally, captive to public transport and non-motorized modes (NMT), while higher income people are more likely to use private motorized transport (Dargay et al., 2007). Furthermore, even in metropolitan areas the overall levels of car ownership are low relative to developed countries (Dargay and Gately, 1999). The South African National Household Travel Survey (NDoT, 2005) found that 42% of respondents used public transport as their primary mode of travel to work, and that 30% either walked or cycled to work. The remaining 28% used private vehicles. These numbers highlight the importance of public transport and NMT to trip making in (South) Africa.

Traditional functional classification, through its application to road access management, dictates the norms around intersection spacing, property access type and spacing, the allowable network interconnects and mode operations along links. If the classifications are not sensitive to the trip making characteristics, particular to the (South) African situation (with its unusually long walking distances and high levels of reliance on NMT and public transport), they will not begin to address the problems particular to South Africa.

The extent to which road categorization influences the provision of transport infrastructure cannot be understated. Road classification, as it is conventionally applied, takes much more from its theoretical underpinnings than is required to produce a generic description of the facility. Instead, it uses this theory as the basis for a more prescriptive interpretation of the function of the road. Conventionally, it concentrates on the traffic functions of streets, and the relationship between the categories developed from the scheme in terms of mobility on the network. The categorization scheme, therefore, generates a hierarchy of levels of mobility that, in turn, is used to develop parametres that define the infrastructural norms for that facility type. The nature of the categorization system, and its interpretation in terms of mobility, means that the primary parametres that are developed are operating speeds and target vehicular traffic volumes. The hierarchical nature of the system is also used to develop norms for determining which class of road can link to which and, also, in general, what quality of service must be provided for which modes and along which links. Furthermore, although motorized transport and NMT often use the same roads, their classification scheme is completely separated from each other, resulting in confusion and a lack of integration of NMT facilities in many areas and, ultimately, an unacceptably high road safety burden.

The road classification approach dovetails very well with other practices in transportation planning, and this is arguably not coincidental, since many of these ideas were being formalized at around the same time. Notably, the concepts behind, and the implementation of, road categorization works very well with the concept of Level of Service (LOS), first formally published in the US in the 1965 edition of the Highway Capacity Manual (HCM) (Highway Research Board, 1965), soon after the publication in the UK of Buchanan's *Traffic in Towns* report.

Alternative categorization schemes

More recently, in an attempt to address some of these concerns, various alternative road categorization schemes have been developed. Notable amongst these is the concept of 'Sustainable Safety' (Institute for Road Safety Research, 2005), developed in The Netherlands, which asserts that a street can only have one function, either mobility or access. This implies that, where a street is designated as a mobility route, all other street functions are actively suppressed, and where it is an access route, mobility is given low priority. To ease the transition between the two extremes, a third 'distributor' category has been introduced (Figure 9.3). The method also advocates the importance of the complete separation of modes along mobility routes, emphasizing that each mode receives the highest LOS practical in any location.

The position taken in the Sustainable Safety approach had been adopted elsewhere as well, e.g. in Australia, Brindle (1995) describes the use of the 'Separate Functions' model to relate the functional mix of roads and streets. The approach described also advocates a sharp distinction between roads whose primary function is to serve traffic, and streets that primarily provide access to properties.

Brindle (1995), however, goes further, arguing that a road's place in the hierarchy should simply be defined by its role in the traffic network. However, both the network role (the estimated traffic on the specific link)

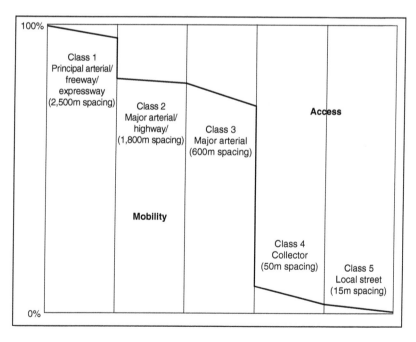

Figure 9.3 COTO road classification system.
Source: COTO (2012).

and the land use and traffic interaction should be used to determine physical characteristics, such as cross-section, geometric design and the treatment of abuttal features. He argues that this differs significantly from the approach that designates road characteristics according to classification. His approach, therefore, limits the influence that the hierarchy has on the final design.

The approach favoured by Brindle (1995) has much in common with the one which is being discussed by the Committee of Transport Officials (COTO) in the new road access management guideline document entitled *South African Road Classification and Access Management Manual (RCAM)* [Committee of Transport Officials (COTO), 2012], which is now applied nationwide. The approach in the COTO document is that roads are classified as being either mobility roads or activity streets.

The approach defines mobility roads as being high speed through routes on which movement is the dominant function, with only limited access to these routes provided by widely spaced intervals; they are described as being 'vehicle priority' or 'vehicle only' routes. Activity streets are defined as catering for the access needs of both vehicles and pedestrians by providing entry to adjacent land uses. These routes are characterized by low operating speeds for safety reasons.

The approach closely mirrors that described by Brindle (1995), in that there is a sharp delineation of function between mobility-focused routes and access-focused routes. The COTO view, however, expands on the principle of separate functions by fitting the current 5-tier classification system to the new classification scheme (see Figure 9.3).

In contrast to these more traditional approaches, the *Arterial Streets for People (ARTISTS)* project (Svensson, 2004), a research project in the European Commission Fifth Framework Programme, completely rethinks the ideas behind classification. It was set up to address a perceived gap in expertise in designing and planning urban arterial streets. The author contends that conventional guidance on the design and management of urban roads (roads being primarily focused on mobility) and streets (being primarily focused on access to properties) has tended to focus on either arterial roads or local access streets. He holds that there is a lack of a clear and consistent approach to the design of arterial streets, which combine both significant through traffic and urban place functions. The project sought to develop a new approach to road planning that addressed these problems.

The approach developed was novel, and was explicitly developed to be 'people-centric', considering the function of the road, both in terms of the specific location along the road being considered, and the route as a whole's place in the network. Svensson (2004) notes that there are often several uses or activities competing for the available urban street-space. These activities may, sometimes, be in conflict with each other, and where they coincide in space they may need to be controlled so that different activities use the same space but at different times. It is the task of street design and management to mediate between competing activities and afford them an appropriate share of space and time.

The novelty of the ARTISTS approach is in the method used to find an appropriate balance between the so-called place-specific functions and the movement function, at any point along the road. The system is based on two unique principles:

1 Any street section has a combination of link status and place status. These are independent of each other, rather than one being the inverse of the other, as with traditional functional classification systems.
2 Link status and place status will depend, not only on the immediate attributes of the street section (including physical form and use), but also on their role with respect to the wider street and urban system, considered as a whole.

The method defines the link status as the relative significance of a street section, as a link in the network. It is based on the link's scale of significance (local access street, district distributor, city arterial) within the network. Care is taken to note that link status may vary by mode, so that a link may be particularly important in the pedestrian network, but not so in the vehicular network, although these may overlap. However, the definition to be applied, in this instance, is actually a restatement of more conventional practice, in that the link status being referred to is actually the 'network function' or 'strategic function' of the link.

In determining the link status, the method acknowledges that the current link status is either predefined by design, or assigned by the relevant authorities, and that these authorities probably remain the most suitable parties to decide this input. However, the report suggests that community inputs should also be sought to understand how road users perceive the strategic importance of the link to their journeys and that this information should be used to supplement expert judgement. Furthermore, this information can, probably, also be gleaned from revealed behaviours or trends in trip making data.

Place status is defined as denoting the relative significance of a street section as an urban place in the whole urban area. A street, or square, may perform a citywide role or a more local role. Therefore, the place status is, like link status, related to geographical scale with regard to frequency and type of use and, in principle, can take on national or international scale significance.

These ideas are then interpreted as a two-dimensional classification scheme, with one axis representing the link status, divided into five categories from local through to national, and the second axis representing the place status of the section being considered, also divided into five categories from local through to national (see Table 9.1).

Envisioning road categorization in this two-dimensional manner has several benefits. The character of different links can be described in terms of its importance to the functioning of the broader road network, without losing

Table 9.1 ARTISTS functional classification

Link status	National	1e	1d	1c	1b	1a
	City	2e	2d	2c	2b	2a
	District	3e	3d	3c	3b	3a
	Neighbourhood	4e	4d	4c	4b	4a
	Local	5e	5d	5c	5b	5a
		Local	Neighbourhood	District	City	National
			Place status			
Legend	Non-arterial streets		Non-arterial roads		Arterial streets	Arterial roads

Source: Svensson (2004).

sight of the importance of various locations and its relative importance in terms of their urban or place status. This allows for a more holistic view of the functioning of the road, by not separating it from the communities and areas it passes through. This aspect of the framework also, implicitly, recognizes that a single link may have varying characters along its length, and that each of these may warrant a different treatment to ensure the optimal overall performance, such performance being measured in terms of more than simply mobility concerns.

The resultant classification table contains 25 different possible combinations of place and network status – each supposedly being a unique description of the overall character of the route. This contrasts starkly with the much more conservative five or six categories of road, defined by the traditional classification schemes discussed before. However, the many distinct categories are both a benefit and a flaw. The subtle variations amongst roads can be better captured and expressed, but the multiplicity of categories can be confusing and, more importantly, as the number of categories increases, the distinctions between categories becomes less meaningful.

The important aspect of the ARTISTS approach to this discussion is that it introduces the concept of the importance of 'place', however loosely defined, into the classification process. Place is clearly an important aspect of any road, since, as discussed by the ARTISTS group, place is a very significant descriptor of the activities conducted at any location, and of the people who are conducting those activities. Also important is the fact that the ARTISTS approach recognizes that 'place' importance varies between different locations, something that is not explicitly included in traditional functional classification. The weakness of the method, however, stems from the subjectivity in describing 'place' importance. A further shortcoming is that route segmentation is equally subjective, and relies upon local knowledge and the intersecting road network to delineate segments.

The role of guidelines in transport planning

Policy and legislation drive the provision of transport infrastructure, in that they set the priorities, strategies and actions to be taken when planning transportation. In order to successfully enact policy, however, guidance is often required to assist practitioners when developing plans. This guidance is most often provided in guidelines developed under the auspices of government and guidelines and, therefore, represents the translation of policy and legislative objectives into practical actions. They advise practitioners as to the best practice with regards to the implementation of these policy objectives and strategies. Guidelines, therefore, play an integral role in the practical interpretation of government policies, being the development of plans and infrastructure and so, ideally, should be revised in light of any policy changes or revisions. If not, practitioners are left to their own devices when interpreting policy, possibly leading to plans and infrastructure that is not in line with policy objectives. The worst-case scenario is that policy changes are simply ignored, because there is no practical guidance as to how to implement it.

The development of guidelines presents an ideal opportunity for a thorough investigation into the practical issues around implementing new policies, and an assessment of where current standard practice may still be appropriate, or not. It provides practitioners with the opportunity to interrogate current thinking and systems, and identify where these must be adapted, or abandoned and replaced, if need be, to achieve policy objectives. Often, a good starting point is to look abroad to see what else has been done, under similar circumstances, which is, to a large extent, how the current range of guidelines originated. This international experience may then be adapted as befits local circumstances, where appropriate, or a requirement for new research into a specific issue may be identified if no suitable solutions are found elsewhere.

In any event, the method used to implement policy, which may be data driven itself, and which forms the content of a guideline, should be grounded in sound, scientific reasoning which, in the field of transportation, is often verified by solid empirical evidence. Most often, research findings from the academic realm, after a period of dissemination, peer review and debate, work their way into policies and, eventually, form the basis for the practical guidance given in guidelines.

It is because of this that deviations from the recommendations of guidelines, despite the fact that these only represent some form of general advice to practitioners, are avoided. The scientific standing, and the general acceptance of the recommendations in guidelines as being the best practice, means that they offer a measure of indemnity from professional or legal challenges to the product produced or the effects it has.

Guidelines, however, are developed with a significant level of generalism. They cannot address, with any degree of specificity, all the possible situations that may be encountered in practice. Their use is subject to the acknowledgement that the practitioner must apply the principles outlined in the guideline, as best as possible, to his unique situation. In fact, there is very seldom a situation where an example, given in a guideline, fits the reality of a particular circumstance exactly. The overly rigorous adherence to the letter of any guideline is, therefore, foolhardy, even more so when the guideline may, in some respects, be outdated.

However, as mentioned, the disincentives to deviating from guidelines are great, as alternative approaches may, for example, require lots of data collection or the procurement of proprietary software. Complicating the issue, many road authorities have developed their own sets of design standards that have been based upon existing guidelines (see SANRAL, 2009; Visser et al., 2003), and these tend to be enforced with even greater compliance requirements. This would not be so much of a problem if the guidelines that are currently in use were regularly revised to reflect the current policy objectives but, unfortunately, many of them are not.

The role of context in road planning and design

According to the US Federal Highway Administration, Context-Sensitive Design (CSD) [interchangeably referred to as Context-Sensitive Solutions (CSS)] is:

> *[A] collaborative, interdisciplinary approach that involves all stakeholders to develop a transportation facility that fits its physical setting and preserves scenic, aesthetic, historic and environmental resources, while maintaining safety and mobility. (FHWA, 2007)*

The New York State Department of Transportation defines CSD as:

> *[A] philosophy wherein safe transportation solutions are designed in harmony with the community. (De Cerreño and Pierson, 2004)*

The principles of CSD, and methods for quantifying and measuring the performance of projects in terms of CSD principles, are outlined in a report (Stamatiadis et al., 2009) compiled for the Transportation Research Board. Amongst other aspects, this approach includes the use of interdisciplinary teams, the need to address all alternative modes, the need to maintain environmental harmony and the need to utilize the full range of design choices.

Defining context remains a challenge, since it refers to all the external influences and factors that play a role in the success of a project – that success, itself, being dependent upon how it is measured. The various sources reviewed, all from the US, include an array of factors, and there does not appear to be consensus on where the line should be drawn.

In Utah, in the US, the state Department of Transportation includes physical, social, economic, political and cultural impacts in its definition, although it does not give specifics as to what these include, or how to assess these, supposedly relying on literature, such as Stamatiadis et al. (2009) to support the application of the policy.

The California Department of Transportation (Caltrans) notes that, in the development of transportation projects, the social, economic and environmental effects of projects must be considered fully, along with technical issues, so that final decisions are made in the best overall public interest (Caltrans, 2005). They mention that attention should be given to the need for safe and efficient transportation, the needs of people with disabilities and other vulnerabilities, and the costs of eliminating or minimizing adverse effects on natural resources, environmental values, public services, aesthetic values and community and individual integrity.

According to Olszak et al. (2008), the Maryland Department of Transportation, FHWA and AASHTO co-sponsored a workshop in 2008 called *Thinking Beyond the Pavement: A National Workshop on Integrating Highway Development with Communities and the Environment*, which refined the definition of the CSD process. The workshop envisioned CSD as follows:

> *Context sensitive design asks questions first about the need and purpose of the transportation project, and then equally addresses safety, mobility, and the preservation of scenic, aesthetic, historic, environmental, and other community values. Context sensitive design involves a collaborative, interdisciplinary approach in which citizens are part of the design team. (Neuman et al., 2002)*

The Minnesota DOT and the Kentucky Transportation Cabinet (KYTC) have developed similar CSD principles. Amongst the key concerns mentioned is the need to balance safety, mobility, community, and environmental goals in all projects, the need to involve the public and affected agencies early and continuously, the need to address all modes of travel and the need to apply flexibility in interpreting and implementing design standards. They also mention the importance of incorporating aesthetics as an integral part of good design.

The emergence of CSD highlights the growing trend amongst transportation planners and designers to see that the impacts of the infrastructure they build are much broader than simply the transportation function they are planned for. What CSD attempts to do is to ensure that road infrastructure is viewed more holistically, as being part of a more complex system, not apart from it. The road, therefore, has to sit comfortably in its environs, managing to fulfil its transport role and still be sensitive to the particularities of the environment it is in, whether that is an urban centre, a suburban township or a rural hinterland. These sentiments are captured in the various philosophies underpinning the practice, such as emphasizing the importance of safety for the user of the road, and the community through which it passes, and stressing the need for harmony with the environmental, aesthetic and historic resource values of the area.

The principles espoused by CSD, however, have proven more difficult to apply practically. Road planning and design is often typified by a mechanistic process, involving the application of established mathematical relationships and engineering norms (in the form of parametres and coefficients) to site-specific data (such as land use and population statistics, traffic counts, geological and hydrological information and geographic data), yielding repeatable results that can be unambiguously interpreted into designs. The philosophies and terminologies of harmony, aesthetics and value are quite alien to the usual practices employed when planning and designing roads.

This emphasis, placed on identifying the inherent flexibility in the laws, regulations and best practices for road planning and design, speaks to the policy of many designers to apply design guidelines as rigidly as possible, using them as minimum standards so as to avoid exposure to tort liability (Jones, 2004). It also, however, highlights the need for an alternative mechanism to harness the flexibility inherent in the guidelines, more effectively, so as to achieve some of the objectives of CSD using the quantitative assessment and evaluation skills already commonplace in the practice.

In a certain sense, the combination of the ARTISTS concepts of 'place status' and 'link status' is similar to the context, since the context of any location along a road is composed of both aspects of the place (activities), and aspects of the link (mobility). The challenge, however, is to define what this means in terms of provision for road functions. Infrastructure should be configured so as to provide the best balance between the needs related to the activities performed, at any location, and the mobility needs along the route. Traditional classification methods attempt to find the best compromise between these needs by prioritizing in relation to mobility (higher order roads allow for higher levels of mobility, lower order roads for less mobility). However, travel is a derived need, driven primarily by the need to conduct activities that are spatially separated and, in many instances, the travel associated with any activity is multi-modal. One might drive from home or use public transport to a commercial district, but the actual shopping is done on foot.

There is, therefore, a link between the context of a location, the activities performed at the location and the mode of travel used for the trip. Each mode used in a road has its own specific characteristics and needs, which determine the design parametres for that mode. Also, each location in a network, or along a road, is defined by a set of contextual parametres that determine how, and by whom, the road, at that location, is most often used. It is when these modal characteristics and location-specific factors, or the needs of a mode and the use of a location, align that an ideal planning solution is found.

Accordingly, certain modes are better suited to a particular set of contextual circumstances than others. Therefore, under a given mix of contextual circumstances, certain modes should be given a higher priority than the rest. Priority, then, and by extension context, can be used to determine infrastructural needs. There is still the question of defining the context, and how this relates to the various modes of transport. Since context is a description of who is using the road, and why they are there, the natural descriptors of context should be found amongst descriptors of land use (such as type and intensity), demographic or socio-economic factors (such as densities, employment levels, income levels and age groups), environmental factors (such as environmentally or culturally sensitive areas) and transportation factors (such as the presence of public transport facilities and the number and type of trips through the area). These are all spatially (and temporally) varying information sets, and so are best analyzed using spatial analysis techniques.

Context-sensitive multi-modal planning

Since the 1990s, the integration of Geographical Information Systems (GIS) and Multi-Criteria Decision Analysis (MCDA) has attracted significant interest as GIS gained in popularity (Power, 2003). GIS is well suited to analysing large and disparate sources of information, and spatial decision problems often involve a large set of feasible alternatives and multiple, sometimes conflicting, evaluation criteria. Accordingly, GIS and MCDA are particularly well suited to each other. GIS techniques and procedures have an important role to play in analysing decision problems, whereas MCDA provides an established set of methodologies for structuring decision problems, and designing, evaluating and prioritizing alternative decisions.

The spatial application of MCDA techniques, SMCA, is being applied to an increasing number of different spatial decision problems, and increasingly, transportation problems too. Typical applications include site selection problems (Carver, 1991; Openshaw et al., 1989) and routing problems (Bailey et al., 2005; Rescia et al., 2006).

Whereas, with other SMCA applications, the alternatives assessed were either different sites or routes, Beukes et al. (2011) demonstrated the use of SMCA where the route was predetermined and, instead, the alternatives assessed were the modes of transport using the route. This has important implications for the results. Previous applications generated a single accumulated map from the combination of the standardized criterion maps of the various spatial indicators and derived routings using a least-cost algorithm. Beukes (2011) generates an accumulated map to describe the contextual suitability for each mode of transport. This results in numerous accumulative maps (one for each mode of transport). In this instance, each map shows the relative suitability of a particular mode of transport to a particular location. Since five modes of transport were used in the analysis, five suitability maps are generated. The criteria used in the assessment are shown in Table 9.2.

Table 9.2 Assessment criteria used in the SMCA

Category	Criteria	Interpretation for NMT
Land use	Land use	Certain land uses are better suited to NMT.
	Household density	Higher density areas are more suited to NMT.
Socio-economic	Employment	Areas with high levels of employment are better suited to NMT.
	Income	In terms of NMT policy, NMT should be encouraged across the city, with specific focus on provision in low-income areas.
	Proportion of vulnerable road users	Higher levels of children and elderly increase the need for NMT facilities.
Environmental	Proximity to heritage sites	Heritage sites should be made more accessible to NMT modes.
	Proximity to wetlands	Environmentally important areas should be made accessible to NMT modes.
	Proximity to ecologically significant areas	Environmentally significant areas should be made accessible to NMT modes.
Transportation	Public transport (PT) demand	NMT access to PT facilities should be improved.
	Private car demand	High volumes of vehicular traffic are dangerous for NMT.
	Proximity to PT stops	NMT access to PT facilities should be improved.

Each criteria included in the assessment is represented spatially, as a map, and assessed in terms of its impact upon the operations of the five modes of transport. For every mode of transport, the aggregate performance, or suitability score, is then generated and also represented spatially. Figure 9.4 maps the aggregated suitability scores for bicycles in Cape Town, with darker areas being more suitable for bicycles.

Beukes (2011) applied the methodology to assessing the suitability of modes of transport along arterials as shown in Figure 9.4. The next step in the method involved extracting the suitability scores along the centreline of the arterials and clustering them to identify regions of contextual similarity.

In theory, contextually similar areas along the route could receive the same design treatment, simplifying the road planning process. The data that was

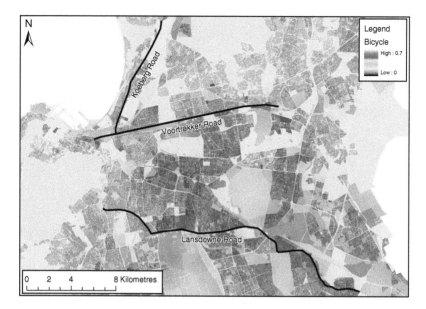

Figure 9.4 Bicycle suitability scores from SMCA for case study routes

clustered consisted of 5-dimensional points with each dimension being the suitability score of a mode of transport at a particular point. The clusters produced, therefore, represented context types, each described by a particular mix of modal suitability.

The cluster means for each cluster within the data set describes the contextual characteristics of each section of the route in a much more compact form than can be achieved by simply analysing the raw SMCA data. There is, of course, some measure of information loss intrinsic to substituting the point scores with the cluster mean scores, since these will invariably differ. But one of the major benefits of clustering is that it makes the contextual assessment of the route more manageable, whilst minimizing information loss. Identifying clusters simplifies further analysis of the data in that much of the complexity, or 'noise', within the data set is eliminated (see Figure 9.5).

However, since the suitability maps were generated for the city, as a whole, this presented the opportunity to conduct an area-wide assessment as well. Raynor (2011) undertook the exercise of clustering the entire city. He used, as input data, the suitability maps, such as in Figure 9.4, generated by Beukes (2011). The hypothesis assumed was that there is an underlying contextual structure across the city, and that this can be used to establish the priorities that should be afforded to each mode of transport, in each area of the city. Although each section of the city will be unique, it is likely that there will be locations that are spatially separate, but also contextually similar, and that these should, therefore, be treated similarly.

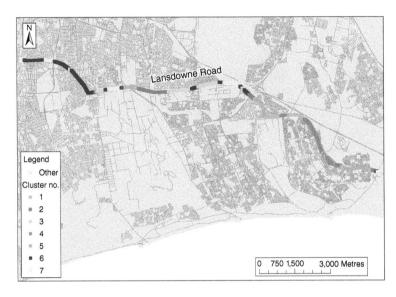

Figure 9.5 Clusters plotted on road network

Cluster descriptions

The work, conducted by Raynor, also involved analysing how clusters form with each additional increment in the number of clusters that was searched for. This gives useful information about the robustness of clusters and, hence, also the relative contextual difference of the areas in those clusters. A cluster that remains stable, irrespective of the number of clusters chosen in the analysis, consists of areas that are very distinct from others outside of its cluster.

Table 9.3 shows how clusters split as additional clusters were added to the analysis. Based on the analysis, the clusters produced in the five-cluster scenario were considered most meaningful, as it captures more detail while not jeopardizing too much significance. As cluster numbers grow, the detail captured increases but the relative significance of clusters diminishes. The five-cluster scenario was found to provide the best balance between detail and cluster significance.

The five-cluster scenario sees the formation of two rural clusters, A1 and A2, and three urban clusters: B1, which is made up only of areas with high public transport (PT) and NMT suitability and low private vehicle and freight vehicle suitability; BC2, which has high suitability scores for all modes, slightly favouring NMT and PT over private and freight vehicles; and C1, which is the polar opposite of BC2, with high suitability for private and freight vehicles.

Table 9.3 Cluster split pattern (PT = public transport)

First split	Cluster	A	B	C			
	Sum d	52,943	36,715	25,522			
	Car	0.3438	0.3817	0.4718			
	Bike	0.3005	0.4288	0.4029			
	Freight	0.2991	0.3560	0.4427			
	Ped	0.2883	0.4458	0.4059			
	PT	0.2789	0.4281	0.4098			
Second split	Cluster	A1	A2	B	C		
	Sum d	17,254	9,310.9	35,961	24,629		
	Car	0.3056	0.4011	0.3823	0.4722		
	Bike	0.3062	0.2931	0.4288	0.4033		
	Freight	0.2632	0.3532	0.3566	0.4432		
	Ped	0.3015	0.2699	0.4458	0.4063		
	PT	0.2840	0.2726	0.4281	0.4104		
Third split	Cluster	A1	A2	B1	BC2	C1	
	Sum d	15,963	15,095	8,990.3	14,440	13,898	
	Car	0.3059	0.3479	0.4011	0.4234	0.4853	
	Bike	0.3054	0.4210	0.2928	0.4305	0.3944	
	Freight	0.2633	0.3196	0.3533	0.3999	0.4537	
	Ped	0.3003	0.4331	0.2695	0.4510	0.3897	
	PT	0.2830	0.4180	0.2722	0.4358	0.3982	
Fourth split	Cluster	A1	A2	B2	BC3	BC4	C2
	Sum d	15,625	6,956.2	8,263.2	12,581	8,923.9	6,890
	Car	0.3062	0.3287	0.4013	0.4878	0.4264	0.4007
	Bike	0.3053	0.4318	0.2917	0.3951	0.4382	0.4033
	Freight	0.2635	0.3051	0.3533	0.4562	0.4037	0.3697
	Ped	0.3003	0.4538	0.2682	0.3903	0.4611	0.4076
	PT	0.2829	0.4332	0.2709	0.3994	0.4477	0.3949

Combining NMT planning and cluster analysis

Given the information from the cluster analysis, it is possible to identify those areas in the city where additional scarce resources could be best applied to improving NMT facilities. In the five-cluster scenario, clusters A1 and A2 represent rural or undeveloped areas and, of the three urban clusters, cluster B1 is made up of areas with high PT and NMT suitability and low personal motorized transport (PMT) suitability; cluster BC2 has high suitability scores for all modes, slightly favouring NMT and PT over PMT; and cluster C1 has high suitability for private and freight vehicles.

In terms of the overarching hypothesis, in general, the transport infrastructure provided should, in any cluster, reflect its contextual character. Thus, NMT facilities should be especially important in clusters B1 and BC2. Overlaying the routes identified in the NMT planning process undertaken by the city municipality with the areas in clusters B1 and BC2 enables routes to be highlighted for priority attention in the network.

The proposed NMT plan was overlaid onto the cluster map for Cape Town, and those sections of the NMT network that fall within these clusters were extracted. A total of 48.19% (or 927 km) of the routes falls within these areas, of which 16.66% (320.5 km) are in cluster B1 areas, and 31.52% (606.5 km) are in cluster BC2 areas.

Given that there are differences between these clusters, in terms of the relative suitability of NMT modes, further distinction can be made on which routes to focus on, in particular. In cluster B1 areas, NMT and public transport modes are much better suited than car and freight modes, whereas in BC2 areas, all modes are more or less equally important. Policy or project objectives could, therefore, dictate which of these areas to focus on, since they both include a substantial length of NMT routes.

In addition, within each of these cluster areas, there are differences in the overall suitability of one or another mode, the cluster mean being only a representative value of the score range in that cluster. Specific projects can, therefore, be identified by focusing on a given mode of transport and looking at where that mode is most suited within a given cluster. The method produces data that can be used to extract areas using suitability thresholds, and routes can be prioritized according to whether they lie within a certain suitability threshold.

In Figure 9.6, a suitability threshold of 75% of the maximum suitability in the cluster B1 areas was extracted and the sections of the NMT network in the cluster B1 areas that lie within this threshold were identified. Of the 320.5 km of NMT routes in cluster B1 areas, approximately 160 km lies within areas where the suitability score for bicycle exceeds 75% of the maximum score for bicycles in cluster B1. Similar techniques can be used to identify areas with high suitability for pedestrians or public transport, and these can be combined to show those routes with high suitability for more than one mode of transport. The example shows the power of using SMCA and cluster analysis to identify project priorities.

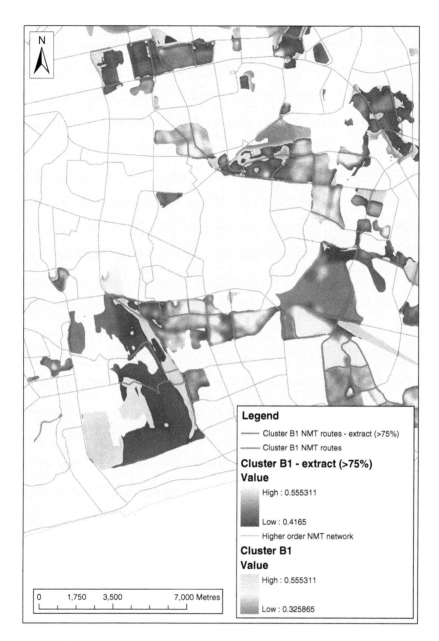

Figure 9.6 Project prioritization approach

Conclusion

The discussion thus far shows that, in general, the conflict between the mobility needs along a road, and the other functions the road must accommodate (however many you choose to recognize), presents road designers with significant challenges. If these challenges are not satisfactorily overcome, the effects can be very serious indeed, leading to infrastructure that is inefficient and unsafe.

Attempts to address the problem have, generally, tried to recast the categorization problem in more rigid terms, describing roads as being either suited to mobility needs only, or access needs only, instead of the more traditional view that many roads actually support a mix of functions.

The more unusual approach, taken by the ARTISTS project, introduces a number of novel ideas to categorization, describing a road segment as being a mixture of place importance (as opposed to the usual definition of access to properties) and link importance. In this way, the ARTISTS approach acknowledges that many roads have varying characters along their length, and that there is a multitude of variations that can be applied to any section of road. This, generally, cannot be accommodated in other categorization schemes, which may even be said to be hostile to the idea.

The emergence of the CSD movement in the US highlights that incorporating local contextual variation into the design process is increasingly being seen as important. However comprehensive it may be, CSD has struggled to gain traction amongst planners and designers, in part because there are no concisely outlined, practically implementable methodologies, and because of concerns around professional liability issues regarding exploiting the flexibility in design standards to meet its principles. Although there are clear benefits to the approach, it has not managed to find support beyond the US, yet.

There are a range of factors that can be used to describe the characteristics of road users, and the activities they are involved in. This information should play a more direct role in the design of these roads. The factors, collectively termed the context, have tended to be overshadowed by concerns around efficiency and cost, and there has not been a comprehensive framework within which the context could be evaluated, and its implications investigated. The way in which infrastructure interfaced with, or suited the context has, thus, always been left to the discretion or judgement of the engineer or planner of the facility.

The chapter identified which factors could be used to describe the context of a location, and demonstrated that it is possible to quantify the context in terms of its effects on the suitability of the various modes of transport. Quantification has a number of advantages for planning and designing infrastructure. Being able to quantify the suitability of a mode of transport to a particular location, given the context, allows for the prioritization of modes in terms of infrastructure provision, which can then be used as the basis for planning and design.

Quantification also has the benefit of facilitating accurate comparisons between different locations along a route. This is useful for planning and design, in that the subtleties in the variation of the context along the route are retained during the analysis, allowing for a fine grained tailoring of the required infrastructure. It is also possible to show that context varies spatially, that it is not static, and that for infrastructure to be contextually sensitive it must, therefore, vary accordingly.

The context, being an amalgam of a range of disparate factors, does not have any intrinsic meaning by itself. Instead, it is the implications of the context that have meaning and, therefore, context can only be understood in terms of its implications for other aspects of the facility. In this research, context is defined in terms of its implications for the suitability of the various modes of transport.

The approach can be used to investigate the context along a single route, or on an area-wide basis, where strategic planning decisions can be informed by the findings of the study. The approach can also be used to optimize and prioritize projects.

References

American Association of State Highway and Transportation Officials (AASHTO) (ed). 2004. *A policy on the geometric design of highways and streets*, 5th Edition. Washington, DC: American Association of State Highway and Transportation Officials (AASHTO).

Bailey, K., Grossardt, T. & Jewell, W., 2005. Participatory routing of electric power transmission lines using the EP-AMIS GIS/Multicriteria Evaluation Methodology. In *Proceedings of the 10th International Symposium on Information and Communication Technologies in Urban and Spatial Planning and Impacts of ICT on Physical Space (CORP 2005)*. Wien, Austria, pp. 137–142.

Behrens, R., 2005. Accommodating walking as a travel mode in South African cities: Towards improved neighbourhood movement network design practices. *Planning Practice and Research*, 20(2), pp. 163–182.

Beukes, E. et al., 2011. Creating Liveable Neighborhoods Through Context-Sensitive Multimodal Road Planning. *Transportation Research Record: Journal of the Transportation Research Board*, 2244, pp. 27–33.

Beukes, E., Vanderschuren, M. and Zuidgeest, M. 2011a. Context sensitive multi-modal road planning: a case study in Cape Town, South Africa. *Journal of Transport Geography*, 19(3): 452–460.

Beukes, E., Vanderschuren, M., Zuidgeest, M. and Brussel, M. 2011b. Creating liveable neighbourhoods through context-sensitive multi-modal road planning. *Transportation Research Record: Journal of the Transportation Research Board*, 2244(1): 27–33.

Brindle, R. (ed). 1997. *Special report 53: Living with traffic*. Vermont South, Victoria, Australia: ARRB Transport Research Ltd.

Buchanan, C. 1963. *Traffic in towns*. London: Penguin Books.

Caltrans 2005. *Main streets: flexibility in design and operations*. Sacramento, CA: California Department of Transportation.

Carver, S., 1991. Integrating multi-criteria evaluation with geographical information systems. *International Journal of Geographical Information Science*, 5(3), pp. 321–339.

Committee of Land Transport Officials (COLTO). 1986. Urban transport guidelines: draft UTG1 - guidelines for the geometric design of urban arterial roads. Pretoria: Council for Scientific and Industrial Research.

Committee of Transport Officials (COTO). 2012. South African road classification and access management manual (TRH 26), version 1.0. Pretoria: Committee of Transport Officials (COTO), Department of Transport.

Dargay, J. and Gately, D. 1999. Income's effect on car and vehicle ownership, worldwide – 1960–2015. *Transportation Research Part A: General*, 33: 101–138.

Dargay, J., Gately, D. and Sommer, M. 2007. Vehicle ownership and income growth, worldwide – 1960–2030. *The Energy Journal*, 28(4): 163–190.

De Cerreño, A. L. C. and Pierson, I. 2004. *Context sensitive solutions in large central cities*. New York: Rudin Center for Transportation Policy and Management.

FHWA. 2007. What is CSS? Accessed at: http://contextsensitivesolutions.org/content/topics/what_is_css/

Forbes, G. 1999. Urban roadway classification. In *Urban Street Symposium*, Dallas, Texas. Washington, DC.

Highway Research Board. 1965. Highway Research Board special report 87: highway capacity manual. Washington, DC: Transportation Research Board.

Iaych, K., Alexeev, V. and Latipov, O. 2009. IRF world road statistics 2009. Geneva: International Road Federation.

Institute for Road Safety Research. 2005. *Advancing sustainable safety*. Leidschendam, Netherlands: Institute for Road Safety Research (SWOV).

Jones, R. 2004. Context-sensitive design: will the vision overcome liability concerns? *Transportation Research Record: Journal of the Transportation Research Board*, 1890(1): 5–15.

Lillebye, E. 1996. Architectural and functional relationships in street planning: an historical view. *Landscape and Urban Planning* 35(2–3): 85–105.

Mabunda, M. M., Swart, L.-A. and Seedat, M. 2008. Magnitude and categories of pedestrian fatalities in South Africa. *Accident Analysis & Prevention*, 40(2): 586–93.

Marshall, S. 2004. Building on Buchanan: evolving road hierarchy for today's streets-oriented design agenda. In *Proceedings of the 2004 European Transport Conference*. London: Association for European Transport, pp. 1–16.

Marshall, S. 2005. *Streets and patterns*. Routledge.

Msemburi, W., Pillay-van Wyk, V., Dorrington, R.E., Neethling, I., Nannan, N., Groenewald, P., Laubscher, R, Joubert, J., Matzopoulos, R., Nicol, E., Nojilana, B., Prinsloo, M., Sithole, N., Somdyala, N. and Bradshaw, D. 2014. Second national burden of disease study for South Africa: cause-of-death profile for South Africa, 1997–2010. Cape Town: South African Medical Research Council, ISBN: 978-1-920618-34-6.

NDoT. 2005. *Technical Report: The First South African National Household Travel Survey 2003*, Pretoria, South Africa: South African National Department of Transport.

NDoT, 2006. *National Road Safety Strategy*, Pretoria, South Africa: South African National Department of Transport.

Neuman, T.R. et al., 2002. *A guide to best practices for achieving context sensitive solutions*, Washington, D.C. pp.2

O'Flaherty, C. 1997. *Transport planning and traffic engineering*. Butterworth-Heinemann.

Olszak, L., Goldbach, R. and Long, J. 2008. Do context-sensitive solutions really work? *Transportation Research Record: Journal of the Transportation Research Board*, 2060, 107–115.

Openshaw, S., Carver, S. & Fernie, J., 1989. *Britain's Nuclear Waste: Siting and Safety*, London, UK: Belhaven Press.

Power, D.J., 2003. A brief history of decision support systems. Retrieved from: http://dssresources.com/history/dsshistory.html. Last accessed on 03/05/12.

Raynor, J. R., 2011. *Context Sensitive Road Planning And City-wide Cluster Analysis*. University of Cape Town.

Rescia, A.J. et al., 2006. Environmental analysis in the selection of alternative corridors in a long-distance linear project: a methodological proposal. *Journal of Environmental Management*, 80(3), 266–78.

SANRAL, 2009. *Design Guidelines for Single Carriageway National Roads*, Pretoria, South Africa: South African National Roads Agency Limited.

Sampson, J. 2010. Road classification and access management: what authorities need to know. In 4th South African Road Federation/International Road Federation Conference for Africa, "Preserving Africa's Road Network". Somerset West, South Africa.

Stamatiadis, N., Adam, K., Don, H., Theodore, H. and Jerry, P. 2009. Quantifying the benefits of context sensitive solutions. Tech. Rep. 642. Washington, DC: National Cooperative Highway Research Program, Transportation Research Board.

Svensson, A.S. (ed). 2004. Arterial streets for people: guidance for planners and decision makers when reconstructing arterial streets. European Commission Fifth Framework Programme.

Visser, D. et al., 2003. GAUTRANS Guidelines for the provision of pedestrian and bicycle facilities on provincial roads in Gauteng. In *22nd Southern African Transport Conference (SATC2003)*. Pretoria, South Africa.

10 Implementation and evaluation of walking buses and cycle trains in Cape Town and Dar es Salaam

Hannibal Bwire, Patrick Muchaka, Roger Behrens and Patrick Chacha

Introduction

School travel planning is an important, but largely neglected, aspect of the local transport planning process in Sub-Saharan African cities. This neglect emanates from past framings of the transport problem in these cities, which resulted in limited travel behaviour data. In Cape Town, for instance, apartheid policies dictated an analytical focus on the daily transportation of labour in and out of cities. This, combined with a focus on the problem of traffic congestion and highway construction in the travel surveys and demand forecasting methods that have dominated transport planning practice, led to a particular scope in travel behaviour analysis. With some exceptions, the travel surveys were limited to motorized trips occurring within the weekday morning peak period when congestion was generally worst. In many instances only trips to work were included. Most analysis of travel needs and behaviour was restricted to either commuting or motorized travel, and travel by children was either omitted entirely or only partially considered. It is only relatively recently that school trips have been a focus in travel surveys – for example in South Africa's National Household Travel Survey (NHTS) in 2003 (DoT, 2005). Several school travel surveys have subsequently been conducted in Cape Town (e.g. Adam, 2003; Fredericks, 2003; Behrens and Van Rensburg 2009).

A study by Muchaka et al. (2011) (n = 1,075) found that in one higher-income neighbourhood of Cape Town (Rondebosch), the majority of school-children (86%) were driven to school. Of these respondents, most (56%) identified distance as the main reason for not walking, followed by fear of criminals (11%) and the number of bags (containing heavy extra-mural equipment) that need to be carried (8%). A further 3% cited inadequate NMT infrastructure (including few or no footpaths, preferred footways or paths along preferred route obscured by vegetation and few or no pedestrian crossings along preferred routes) as reasons for not walking to school.

Available historical data in Cape Town suggests that there has been a significant shift in mode use for school trips in the higher-income neighbourhoods over the past three decades, with significantly fewer children now using the NMT modes of walking and cycling. For example, a survey conducted in 1976, as part of the Cape Metropolitan Transportation Study, found that amongst 1,020 middle- and high-income households living in Cape Town, 49% of trips to school were on foot or by bicycle, 13% were by train or bus, and 38% were by car. A later survey of 100 households by Market and Opinion Surveys in 1992 suggested that, amongst the same group, school trips by foot or bicycle had dropped to 38%, trips by public transport had dropped to 9%, and trips by car had risen to 52%. A survey of 1,494 pupils conducted by the Centre for Transport Studies in 2009 found that, amongst nine participating schools in Rondebosch, which serve predominantly middle- and high-income communities, trips to school by foot or bicycle had declined to 8% (7% on foot and 1% by bicycle), trips by public transport had declined to 3%, and trips by car had increased to 87% (Behrens and Van Rensburg, 2009). On the other hand, the lower-income neighbourhoods have not experienced a similar shift in mode use for school trips, with the majority of learners still walking to school.

There is also a paucity of knowledge about child mobility in Dar es Salaam. Previous studies have paid little attention to how children move around in public outdoor environments. It is only recently that school trips and other trips have become a focus in travel surveys, for example Bwire (2009) and Bwire and Chacha (2011). This latter study found that the majority of children walk to school (62% and 66% to and from, respectively). The same study found that 81% of parents were worried about the risk of their children being involved in road traffic crashes (60% very worried and 21% quite worried). In addition, the study also indicated that traffic danger was most commonly identified as a reason for child accompaniment by parents to school.

A collation of available historical data, in Cape Town and Dar es Salaam, on education trip NMT main mode share is presented in Figure 10.1.

Despite this paucity of data, the limited available data shows that, despite rapid growth in car use in middle- and high-income households, scholars of low- and middle-income households are heavily dependent on non-motorized modes as their primary means of transport. Available road safety data indicates that children are the most vulnerable road users to road crashes. Media reports in the case cities, particularly Cape Town, indicate that child pedestrians are also vulnerable to crime and molestation. As a result, there is a need for interventions that simultaneously address rising car use among children and concerns over the safety and security of child pedestrians.

Two NMT interventions popular in the developed world – walking buses and cycle trains – have the potential to address each of the above concerns. A walking bus is a group of children who walk to and from school under the supervision of adult volunteers, one of whom leads at the front (the 'driver') and another who supervises at the back (the 'conductor'). Children are picked up either from their homes or from designated 'bus stops' along a set route and dropped off at school. The concept is attributed to David Engwicht (1993), with the first walking bus trialled in the United Kingdom in 1998. Since then, walking buses have been introduced in many other parts of the developed world (e.g. Australia, New Zealand and the US).

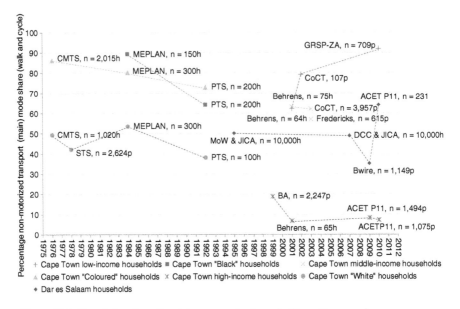

Figure 10.1 Trends in education trip car mode share in Cape Town and
Dar es Salaam (1975–2011).

A cycle train operates along the same lines, using bicycles instead of walking. Cycle trains are well established in Belgium and UK and New Zealand (O'Fallon, 2007).

The increasing international popularity of these interventions stems from their ability to incorporate greater physical activity into children's lives and to reduce traffic congestion within school precincts. Furthermore, adult supervision can ensure a safer journey to and from school, and child participation can facilitate the development of safe road use behaviours and build essential skills that can be used later when walking or cycling independently. Both interventions can provide children and parents with opportunities to build friendships, thereby generating stronger and more liveable local communities (Collins and Kearns, 2005). However, while offering these benefits, these interventions have been criticized as simply replacing one form of adult accompaniment with another, i.e. from chauffeuring to adult supervision (Kearns and Collins, 2003; Hillman, 2006).

Because walking buses and cycle trains are new to the developing world, little is known as to how they can be replicated in a developing country context and with what results. This chapter describes examples of such interventions as they were implemented in Cape Town and Dar es Salaam, and evaluates their impacts. The projects that were demonstrated in Cape Town and Dar es Salaam are described in terms of the implementation method, programme evaluation and programme impacts.

Walking bus interventions in Cape Town

Two walking bus projects were implemented, one in a lower-income neighbourhood (Delft) and the other in a higher-income neighbourhood (Rondebosch). The Delft project was implemented and coordinated by the Global Road Safety Partnership-South Africa (GRSP-ZA), while the Rondebosch project was conceptualized, implemented and coordinated by ACET researchers in the Centre for Transport Studies at the University of Cape Town.

Preparation surveys

Ahead of the implementation phase, two school travel surveys were conducted in the study neighbourhoods (Rondebosch, n = 1,075; Delft, n = 709) in 2010. These surveys served to gain insights into current learner travel behaviour and to gauge parents' interest in walking buses, in addition to creating a database of potential participants. The overall questionnaire response rate was 52% in Rondebosch schools (Rondebosch survey), and 31% in Delft schools (Delft survey).

The pen-and-paper self-completion surveys were divided into two sections. The first was for completion by learners with a parent or guardian's help, and covered learner demographics, travel time to school, mode used to and from school, reasons for not walking among learners who use modes other than walking, and problems faced by learners who currently walk to school. The second section was for completion by a parent or guardian, to elicit parent or guardian attitudes towards walking buses and to collect the contact details of parents willing to either let their children participate or to supervise such buses. The surveys were translated in Afrikaans for the benefit of the Delft participants.

Figure 10.2 shows details of participating schools. The Rondebosch survey participants were drawn from a sample of learners from Reception grade (grade R) to grade 7 at six primary schools that had shown the greatest interest in walking buses in a 2009 feasibility survey (see Behrens and Van Rensburg, 2009). More primary school children are likely than secondary school learners to live within walking and cycling distances of school, and walking buses are only suitable for children at primary school level where parents or guardians play a prominent role in child travel decisions. The interest in promoting NMT for school travel in Rondebosch emanated from a shared concern among parents, City of Cape Town officials and residents' associations in Rondebosch about peak period traffic congestion and child pedestrian and cyclist safety.

In Delft, the interest in promoting safer walking to school was largely based on concerns about the risk of road crashes. Here, most learners between the ages of 5–12 years walk to school, often unaccompanied. This age group is particularly vulnerable, due to their physical and cognitive limitations. Furthermore, Delft is located close to a freeway (R300). The process started with a preliminary set of interviews conducted with 16 of the 32 schools located in and

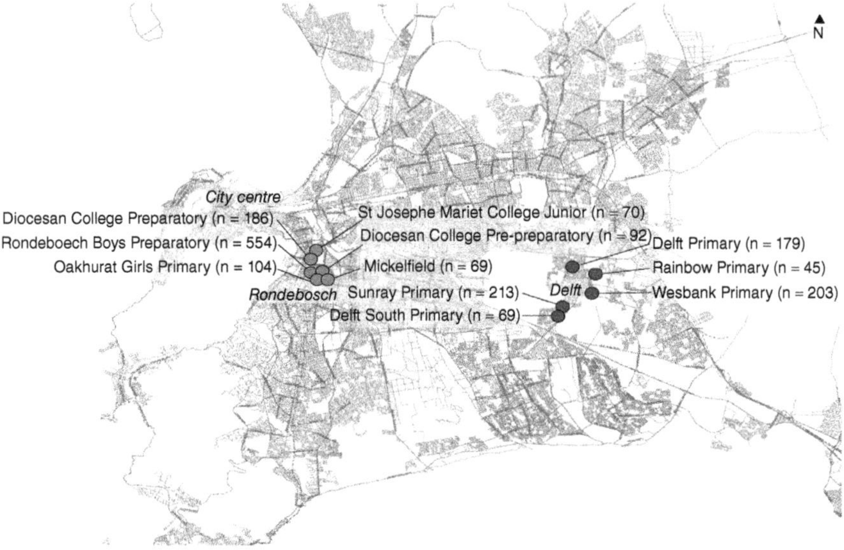

Figure 10.2 Location of participating schools in Cape Town (and sample sizes).

around Delft, among learners from grades 1 to 3, on the basis that children in these grades are more vulnerable to road crashes because of their age, compared to older children in the higher grades. Following the preliminary interviews, five schools in Delft were selected for the more detailed survey, based on their proximity to the R300 freeway and the associated pedestrian safety risk.

Analysis of survey data

Data was analyzed in two phases: an initial phase aimed at extracting information that could be used to set up walking buses (names, contact details and addresses of those willing to join walking buses), and a second phase aimed at more detailed analysis of current travel patterns (why some learners do not walk to school, the problems faced by learner pedestrians, and the reasons some parents were unwilling to participate in the initiative).[1]

Details of participating schools, sample sizes and response rates are shown in Table 10.1. Eighty percent of respondents in Rondebosch schools were male, and 20% female; the higher proportion of male respondents' in the Rondebosch survey resulted from the fact that three of the schools (Rondebosch Boys Preparatory, Diocesan College Preparatory and Diocesan College Pre-Preparatory) are boys' only schools. These schools have a higher combined enrolment total compared to the two girls' only schools (Micklefield Girls Primary and Oakhurst Girls Primary) and the one co-ed school (St Joseph's Marist College Junior).

Table 10.1 Sample size and response rate, by neighbourhood and school (Rondebosch and Delft surveys)

School		Total learners in school	Number of questionnaires distributed in school		Number of questionnaires returned	
				% of school		% returned
Delft[1]	Delft Primary School	1,214	499	41.1	179	35.9
	Delft South Primary School	1,012	437	43.2	69	15.8
	Rainbow Primary School	1,151	430	37.4	45	10.5
	Sunray Primary School	1,222	587	48.0	213	36.3
	Wesbank Primary School	1,317	319	24.2	203	63.6
	Sub-total	**5,914**	**2,272**	38.4	**709**	31.2
Rondebosch	Diocesan College Preparatory School	373	373	100.0	186	49.9
	Diocesan College Pre-Prep. School	212	212	100.0	92	43.4
	Mickelfield School	215	215	100.0	69	32.1
	Oakhurst Girls Primary School	220	220	100.0	104	47.3
	Rondebosch Boys Preparatory School	732	732	100.0	554	75.7
	St Joseph's Marist College Jun. School	300	300	100.0	70	23.3
	Sub-total	**2,052**	**2,052**	100.0	**1,075**	52.4

Note: Only the first three grades of the Delft schools were included in the survey – representing between 26% and 51% of the total learners in these schools

In Delft, 43% were male and 55% female (with 2% item non-response). Respondents' ages ranged between 4 and 14 years.

Data limitations

The results of analysis are not to be regarded as fully representative of all learners and parents, nor of the participating schools in the two study neighbourhoods, nor of all schools in the city, for the following reasons.

Firstly, in Rondebosch, the schools surveyed were only those that showed the greatest parental interest in walking buses during the 2009 feasibility study (see Behrens and Van Rensburg, 2009). It is possible that this introduced bias in favour of NMT use in the results obtained. Had all schools been surveyed, including those that showed lower levels of interest in the initiative, in the feasibility survey, the overall levels of parental interest in the initiative is likely to have been lower.

Secondly, the school group response rates of 52% and 31% would have introduced bias. More specifically, it is probable that parents and learners who were more predisposed to walking to school, or lived within walking catchments, were more likely to respond. The actual statistical indicator of all learners' and parents' support of the concept lies somewhere between the indicator for the responder group, and the same indicator calculated on the assumption that all non-responders were, by definition, not supportive of the initiative in question.

Thirdly, in Delft, the choice of only grade 1–3 may have introduced some bias, as it is possible that the attitudes toward learner NMT use among parents of more vulnerable children in lower grades are different to those of parents of less vulnerable children in higher grades.

Implementation

The survey results suggested that there was sufficient interest among parents and children in the two neighbourhoods concerned to make walking buses a viable intervention. Table 10.2 presents findings on parental interest in walking buses in the two neighbourhoods. For comparative purposes, the results from the 2009 feasibility study are also presented. Two schools in Rondebosch (Rondebosch Boys Preparatory School and Oakhurst Girls Primary School) and one school in Delft (Delft Primary School) were selected as the demonstration schools, on the basis of levels of interest in the initiative shown in the survey results, and proximity of potential participating households (obtained from mapping the addresses given by parents in the surveys). The Delft walking buses started operating in November 2010, while the Rondebosch walking buses started in April 2011.

In Delft, parents were ultimately not recruited as adult supervisors, even though potential parent supervisors had been identified during the survey.

Table 10.2 Parent permission for learner participation in, and parent willingness to supervise, walking buses, by school group (percentage) in Cape Town

		Yes	No	Item non-response	Recording error	Total
Delft, 2010 (n = 709)	Permit child to join walking bus	50.5	29.8	19.7	0.0	100
	Volunteer to supervise walking bus	15.5	57.3	27.2	0.0	100
Rondebosch, 2010 (n = 1,075)	Permit child to join walking bus	41.1	57.3	1.4	0.2	100
	Volunteer to supervise walking bus	17.2	77.2	5.6	0.0	100
Rondebosch, 2009 (n = 1,494)	Permit child to join walking bus	52.5	43.1	4.3	0.0	100
	Volunteer to supervise walking bus	33.1	62.7	4.3	0.0	100

Instead, volunteers from Red Cross (South Africa) escorted children to school. In Rondebosch, parent volunteers supervised the walking buses. In both neighbourhoods, the walking buses used bus stops as collection points rather than collecting children from their homes, and the buses operated in the morning only. In Rondebosch, the same walking bus was used by children from both schools, as these schools are in close proximity to each other. Four routes operated at the two Rondebosch schools, while the Delft school had three routes. To enhance visibility to motorists, participants in both neighbourhoods wore reflective vests and sashes. Selected characteristics of the Rondebosch walking buses are presented in Table 10.3.

Evaluation

As part of its coordination role, ACET researchers undertook an evaluation of the Rondebosch walking buses to assess the impacts and long-term viability. Quantitative 'before' physical activity, odometre and school gate congestion data had been collected in March and April 2011, at the two demonstration schools and at a control school (Grove Primary School in Claremont, Cape Town). The 'after' data was to be collected later in the year after the launch of the buses. However, by the time this 'after' data was to be collected in late 2011, all four buses had ceased operating.

Table 10.3 Selected characteristics of the Rondebosch walking buses

	Number of stops	Route length (km)	Start time	Number of children at launch	Number of parent volunteers	Days per week
Park Road route	2	1.80	07h15	10	6	2
Ave de Mist route	3	1.75	07h15	19	10	5
Keurboom Road route	2	1.45	07h25	6	5	2
Liesbeek Parkway route	2	1.75	07h15	9	4	2

The discontinuation of the walking buses necessitated a methodological switch to qualitative interviews only. The participant interviews (with parents and learners) were conducted between October and November 2011, eight to nine months after the launch of the walking buses. The interviews were preceded by email and telephone requests to all parents who had been involved in the initiative, either as consent-givers or volunteers. Of these, 14 parents agreed to be interviewed. Home interviews were conducted with parents and their children, except in two instances where the children were not present. Of the 14 parents who were interviewed, 11 were mothers and three were fathers, while of the 16 children interviewed, five were girls and 11 were boys. The interviews were conducted using an open-ended question schedule.

Key findings were analysed in terms of learner travel behaviour prior to, and after, the setting up of walking buses; reasons for discontinuing their respective walking bus; and insights gained into the impacts of walking buses from both child and parental perspectives. Matters related to mode use, reasons for stopping the use of walking buses and experiences while using the walking bus are highlighted using selected quotations.

Notwithstanding the small sample, the results do indicate that walking buses have the potential to generate transport system benefits in terms of influencing mode choice change. Of the 12 parents who indicated that, before the walking bus initiative, they chauffeured their children to school every day by car, seven (58%) reported that (at the time of the interview) they now walk with their children to school at least once a week. One parent indicated that her son was now using a lift club only because the walking bus had ceased to operate, and that her child would have continued walking if the bus had continued. Nine (75%) out of the 12 parent respondents were willing to let their children use a walking bus if the service was resumed.

The ability of the walking bus to influence behaviour was described by one parent this way:

> Parent respondent 10 [mother of a 10-year-old boy in grade 4]: *"Prior to the introduction of the walking bus, it had never occurred to me that walking to school could be an option ..."*

Several reasons were given by parents for stopping participation in the walking bus, with four (29%) out of the 14 parents saying it was difficult to walk in winter. It seems the problem with winter was not cold and wet weather *per se*, although this is likely to have played some role. Instead, it was reported that it was too dark to walk in the mornings to feel safe or be seen clearly by vehicle drivers, as illustrated by the following:

> Parent respondent 2 [mother of an 8-year-old boy in grade 3]: *"We stopped when winter set in. Not so much because of adverse weather but because, in winter, it was still too dark around 07h15 when we were supposed to start walking. We hoped to start again in summer. However, the bus is no longer operating so we walk alone on certain days."*

Another reason for the demise of the initiative centred on parent volunteers. One parent said their bus was too small and had too few reliable parent volunteers, resulting in her walking the children to school every day. Although parent respondent 12 cited distance to school (too far) as the main reason for discontinuation, it seems her decision to stop also arose from her frustration with the lack of co-operation from her fellow volunteers:

> Parent respondent 12 [mother of a 10-year-old boy and a 13-year-old girl in grade 4 and 7, respectively]: *"It (the walking bus) enabled me to meet with other parents. However, at times I felt that other parents were not playing their part. I ended up supervising the bus even on days when it wasn't my turn to do so. I would get to the bus stop only to find that the parent on the roster was not there to collect the children. In such cases, I was forced to walk with the children to school."*

Parents and children both appeared to think that the walking buses had a positive impact on children. The majority of children interviewed seem to have enjoyed their experiences with the buses. While some of the parents were no longer interested in the concept, all but one of the children were willing to use it again. The positive impacts were that it was fun and helped them to get some physical exercise, as illustrated by the following:

> Child respondent 1 [7-year-old girl in grade 2]: *"It helps you to get energy out. At times you wake up with lots of energy and you can get that energy out while walking to school."*

Child respondent 2 [8-year-old boy in grade 3]: *"The walking bus keeps you fit and it is a lot of fun."*

Child respondent 8 [10-year-old boy in grade 4]: *"I liked it because we could take the dogs with us on the walking bus."*

Parents' views on the impacts on children were also largely positive, and these also centred on it being fun and allowing children to exercise:

Parent respondent 4 [mother of 10- and 8-year-old girls in grade 5 and 2]: *"The exercise is brilliant for the children as they have very little time to exercise during the day."*

Parent respondent 7 [mother of 13-year-old girl in grade 7]: *"While we have always walked alone, she found it fun walking in a group."*

Parent respondent 9 [father of a 10-year-old boy in grade 5]: *"It is a way for him to burn off energy and meet with other children. However, the bags are a problem considering the distance to school. This might not be an issue if trolleys were introduced to make it easier for the children to walk."*

Besides being fun and a form of exercise, other parents felt the benefits were more to do with preparing children to move around independently:

Parent respondent 6 [mother of 10-year-old boys in grade 4]: *"It is empowering for the children. They can now walk from school alone if I am unable to pick them up. Before the bus, this is something I would have never let them do."*

Parent respondent 8 [mother of 10- and 8-year-old boys in grade 4 and 2, respectively]: *"My children now know how to get home on their own. They use the walking bus route to travel home on their own on some days."*

Two (14%) of the parents felt that the walking bus had had no impact on their children. In the case of one parent (parent respondent 1), this was because the child had been walking to school before the initiative started. In the case of the other parent (parent respondent 5), this was because the child used the walking bus for a very short period of time (one month, and only on Fridays, when he did not have to carry sports equipment).

The impacts reported by parents on themselves were mostly positive, and centred around the walking bus helping to build a sense of community and allowing parents to get to know each other. This is reflected in the following response:

Parent respondent 3 [mother of an 11-year-old boy in grade 5]: *"It has allowed me to get to know parents I may never have got to know. I used to see some of the parents taking their children to school but we did not know each other. Now we know each other and we stop to chat when we meet ..."*

However, there were also some negative impacts which centred on the loss of family time, inflexible schedule and lack of cooperation from other parents:

> Parent respondent 1 [mother of a 7-year-old girl in grade 2]: *"The bus schedule was too inflexible and was impacting on family time. Therefore, I would rather walk alone with my child to school."*
>
> Parent respondent 5 [father of a 10-year-old boy in grade 5]: *"I prefer to drive my son to school as this gives us time to bond with each other."*
>
> Parent respondent 12 [mother of a 13-year-old boy and a 10-year-old girl in grade 7 and 4, respectively]: *"... at times I felt that other parents were not playing their part. I ended up supervising the bus even on days when it wasn't my turn to do so. I would get at the bus stop and find that the parent on the roster was not there to collect the children. In such cases I was forced to walk with the children to school."*

Walking and cycle train demonstration in Dar es Salaam

In Dar es Salaam both walking buses and cycle trains were implemented. These were coordinated by the two authors who are based at the University of Dar es Salaam (Hannibal Bwire and Patrick Chacha).

Preparation surveys

As part of the preparation for launching the walking buses and cycle trains, school travel surveys were conducted at 10 primary and secondary schools across three municipalities (Kinondoni, Ilala and Temeke). These schools were selected because they already had the highest number of children who walk and cycle to/from school or were schools that had received cycling and road safety training from the Dar es Salaam Cycling Community, UWABA (a Kiswahili acronym for this community) and a New York-based non-profit organization, AMEND Tanzania. The schools that took part in the surveys, and their respective sample sizes, are shown in Figure 10.3.

Letters outlining the study were sent to municipal directors, requesting permission to carry out the surveys in 10 selected schools. Head teachers were then visited and asked to nominate a teacher and a random selection of children to participate in the study. The surveys were aimed at gaining insights into current learner travel behaviour and to collect the data required to set up walking buses and cycle trains. The questionnaires were translated and back-translated into Kiswahili.

Following the recruitment of schools, the Dar es Salaam survey (n = 1,511) was conducted between September and November 2011. The pen and paper self-completion questionnaire was largely similar to that used at the Cape Town schools, with additional questions regarding cycle trains. A total of

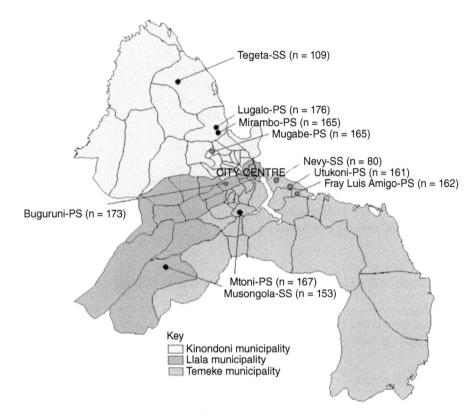

Tegeta-SS (n = 109)

Lugalo-PS (n = 176)
Mirambo-PS (n = 165)
Mugabe-PS (n = 165)

Nevy-SS (n = 80)
Utukoni-PS (n = 161)
Fray Luis Amigo-PS (n = 162)

CITY CENTRE

Buguruni-PS (n = 173)

Mtoni-PS (n = 167)
Musongola-SS (n = 153)

Key
☐ Kinondoni municipality
▨ Llala municipality
▧ Temeke municipality

Figure 10.3 Location of participating schools in Dar es Salaam City
(and sample sizes)

2,000 questionnaires were distributed, and 1,511 were returned, representing
a response rate of 75.6%.

Implementation: walking buses

Parents/adults exhibited a rather low level of interest: about 33% of parents/
adults indicated a willingness to volunteer to conduct a walking bus, once or
twice a week, or 'in two weeks' time'.

The addresses of interested children and adult volunteers were manually
located on the map, in order to show where there was sufficient interest to
establish a walking bus, and what route a walking bus might follow. Once the
route design was completed, participants were sorted into a series of route
groups indicating child participants, their nearby bus stop, and their anticipated usage of the walking bus, as well as the adult volunteers' names, phone
numbers, and possible 'driving' times. Results pertaining to parental interest
in walking buses, from the implementation survey, are presented in Table 10.5.

Table 10.4 Sample size and response rate, by neighbourhood and school
(Dar es Salaam survey)

	Total number of learners in school	Number of questionnaires distributed in school		Number of questionnaires returned	
			% of school		% returned
Lugalo Primary School	924	200	21.6	176	88.0
Mugabe Primary School	667	200	30.0	165	82.5
Buguruni Primary School	2,320	200	8.6	173	86.5
Mtoni Primary School	2,572	200	7.8	167	83.5
Mirambo Primary School	830	200	24.1	165	82.5
Ufukoni Primary School	1,603	200	12.5	161	80.5
Fray Luis Amigo Primary School	568	200	35.2	162	81.0
Nevy Secondary School	756	200	26.5	80	40.0
Tegeta Secondary School	1,080	200	18.5	109	54.5
Msongola Secondary School	456	200	43.9	153	76.5
Total	11,776	**2,000**	17	**1,511**	75.6

Table 10.5 Parent permission for children participation in, and parent willingness to supervise, walking buses, by school group (percentage) in Dar es Salaam City

	Yes	No	Item non-response	Total
Permit child to join a walking bus	49.8	22.2	27.9	100
Volunteer to supervise a walking bus	33.2	20.64	46.1	100

Walking buses were then only established in primary schools, as a high number (60.8%) of children in secondary schools were not willing to participate. The first walking buses in Dar es Salaam were implemented in November 2011, at Lugalo Primary School, with two walking buses and 20 pupils. The number increased to 25 after two days. The buses continued to operate until schools closed in December 2011, and after re-opening of the school in January 2012. The number of walking bus routes also increased. The initiative was later extended to two more schools (Mirambo and Buguruni Primary Schools) in the city.

Because so few adults were willing to supervise the walking buses, the ACET researcher (who was the coordinator) conducted a short study which found that older children who were involved in the walking bus could be used as 'conductors'.

Children and the supervisors (drivers) wore high-visibility sashes and clothes. These served a dual purpose: increasing the visibility of the children and adults to passing cars; and creating a higher profile of the initiative in the surrounding community.

Evaluation: walking buses

Interviews with a sample (parents n = 36 and children n = 36) found that the majority of children enjoyed their experiences with the buses, and agreed to continue using them. Most child respondents said they felt safer crossing the road as a group. One child said that she liked having company while walking to school. Other children said that walking to school provide good physical exercise. Further, children liked the walking bus because it was free and safer than public buses (*daladala*).

All the children noted that *daladala* drivers tend to ignore them when they cross the road, and that cyclists ride on the sidewalks or walkways, which is dangerous to the walking bus users. Another reason for some children disliking the walking bus was the long distances they had to walk.

In contrast to the children, more parents reported negative impacts. Although some saw the buses as a fun, safe and comfortable way for children to reach home/school, one parent said that his children became tired when using the walking bus, and that they wasted much of their time on the road. Another said that the increased tiredness led to a lack of concentration and low performance in the classroom. A further parent said that the absence of dedicated routes, for children walking to Lugalo Primary School, made walking bus use unsafe.

Implementation: cycle trains

Survey results showed that 68% of parents were willing to let their children join a cycle train. In addition, 15.9% were willing to conduct a cycle train, while 8.5% of families were unsure how often they would be able to conduct a cycle train.

The majority of volunteers (73.9%) did not indicate how often they would be willing to conduct a cycle train. The parents who were willing to supervise the trains offered to drive the train one morning or afternoon per week, or every two weeks. This indicated low levels of involvement from parents in conducting cycle trains. Further, the majority of children who participated in the survey (86%) did not own a bicycle. The parents of these children were asked if they would like their child to have a bicycle, and most (79%) responded 'yes'.

The first cycle train was implemented in November 2011 at Fray Luis Amigo Primary School, with 11 learners. Parents were ultimately not recruited as adult supervisors, and instead volunteers from the Dar es Salaam Cycling Community in Dar es Salaam (UWABA) were used as cycle train supervisors. The UWABA volunteers were already experts in cycling and so were familiar with how to handle children on the road and cycleway; together with the researcher, UWABA worked consistently to promote the concept of cycle trains.

Evaluation: cycle trains

A sample (parents n = 15 and children n = 15) of cycle train users were interviewed as part of the evaluation process. This evaluation revealed the following positive impacts.

> One girl said, *"there is good security while using cycle train, children cannot be stolen their bicycles"*. A second girl added, *"cycle trains will reduce bicycle theft"*.
> A parent of a 9-year-old boy stated that *"cycle trains increases bicycle security during cycling to/from school"*. Other parents stated that cycle trains help their children to be more careful when on the road.

However, there were also some negative impacts, which centred on the inflexible schedule:

> Parent 08 responded, *"Sometimes my child comes home late due to waiting for others in the train"*, while parent 17 stated that *"the cycle train is difficult to control"*.

Children were asked if they would like to continue using this mode. Twelve out of 15 (80%) said they would like to do so, as a cycle train protects them from bicycle theft (27%); and because when bicycles break down they are easier to repair with a group (47%). Other reasons for liking the cycle train included better safety due to supervision, and the opportunity for physical exercise.

The three children who said they would not continue using the cycle train were dissatisfied with the use of older children as supervisors, which suggests that, if other supervisors were used, they might continue using the cycle train. A further reason was departure delays on the trip from school, because some of the children received extra schoolwork.

Overall, the cycle trains allowed children who already owned a bicycle, and who knew how to ride, to participate and to improve their skills in a supervised, supportive environment. Cycle trains thus also provide an important developmental stage before independent cycling. If children can be trained as confident cyclists both on- and off-road, it provides them with greater choices for their travel as they go through life. By going through the various stages of cycling, children became more confident about cycling alone.

Cycle trains are also more suitable for children who live further away, at a distance where walking becomes less of an option. Within the wider community, cycle trains helped to raise awareness of cycling, and to raise the profile of children cycling on the roads. Increased awareness and acceptance of cycling in the general population will facilitate its growth as a transport mode, with beneficial implications for issues such as health and traffic congestion in Dar es Salaam city.

Conclusion

One of the key lessons from the research experience and international precedent, in the case of Cape Town, is that the institutional arrangements surrounding walking buses, and the degree of proactive support provided by the school and the local municipality, is just as important as the technical questions around setting up the walking buses and optimizing routes and schedules. In retrospect, insufficient attention was given to establishing these institutional arrangements. This, partly, explains why the Cape Town walking buses were short-lived (less than 6 months). While perhaps briefer than most, this short life span, however, mirrors findings from previous experiences elsewhere. For instance, in Christchurch (New Zealand), Kingham and Ussher (2005) found that 26 out of 56 routes that started operating in September 2000 (at the start of the initiative) had ceased operating by mid-2003 (at the time of the study). Similarly, in the United Kingdom, Mackett et al. (2005) reported that 12 out the 26 buses included in their study in Hertfordshire County, which began in January 2002, had ceased operating by the time of their evaluation survey in May 2002. This suggests that, while walking bus programmes supported by local authorities may last for several years, individual walking bus routes tend to have much shorter life spans.

The implication for local municipalities, interested in promoting NMT initiatives for school travel, is that they should be directly involved in such initiatives in the following key areas: promoting the adoption of initiatives at schools; risk assessment of routes and making any necessary engineering improvements; providing funds to cover costs (e.g. acquiring trolleys for school bags, and reflective vests and training volunteers); and actively supporting the initiative after setup by promoting it and providing ongoing incentives. For schools interested in promoting NMT initiatives for school travel, the implications are that they should also be more involved. While the international literature suggest that schools are typically not directly involved in the day-to-day running of 'walking buses' (see, for example, Mackett et al., 2003; Kingham and Ussher, 2005), they can help significantly in setting them up and sustaining them. Involvement of the school can take the form of ongoing promotion in their newsletters and parent meetings, and through the facilitation of ongoing recruitment of parent volunteers and of the establishment and maintenance of parent organizing committees.

With regard to Dar es Salaam, a key lesson from the research experience is, when implementing initiatives such as walking buses and cycle trains, there are different options with regard to supervisors. As a result, insufficient numbers of parent volunteers need not stop the implementation of these initiatives. The use of volunteers from non-governmental organizations is a model that may be implemented elsewhere when there is sufficient parental interest in letting children join walking buses or cycle trains, but parents are not available to supervise. However, to make this model of supervision more sustainable, funds (e.g. from the relevant local authorities/municipalities) may have to be made available in order to pay some form of allowance to the volunteers who are not parents of participating children. The case of Dar es Salaam supports this idea, as the volunteers requested to be paid in order for them to continue supervising walking school buses and cycle trains. With regard to the use of older children as supervisors, the Dar es Salaam case study is, arguably, the only example where older children, rather than adults, have acted as supervisors of walking buses.

In summary, walking buses are appropriate for both lower-income and higher-income neighbourhoods. The institutional arrangements surrounding walking buses, and the degree of proactive support provided by the school and the local municipality, is as important as the technical aspects concerning setting up the walking buses and optimizing routes and schedules. Non-governmental organization volunteers, and older schoolchildren, can be used to supervise cycle trains and walking buses. These initiatives have had some broader influence in the case study cities and, more particularly, in Cape Town, where, following the demonstration, other organizations in South Africa have started their own initiatives. The main challenge is how to make the walking bus and cycle train initiatives sustainable through their adoption by schools as a part of permanent school programmes, and by non-governmental organizations as permanent projects.

Note

1   For a full discussion on the results of the walking school bus implementation surveys conducted in Cape Town, see Muchaka et al. (2011).

References

Adam, F. 2003. *Development and administration of a pilot learner travel survey*. Final year Civil Engineering undergraduate thesis, University of Cape Town.

Arrive Alive. 2012. Scholar patrols. Accessed at: www.arrivealive.co.za/pages.aspx?nc=Scholar_Patrol_Implementation (accessed 17 July 2012).

Behrens, R. and Van Rensburg, J. 2009. Key results of a feasibility study of non-motorised travel initiatives amongst selected Rondebosch schools. Rondebosch Schools Non-motorised Transport Initiative, ACET Project 11: School Travel Planning. Working paper 11–01, University of Cape Town.

Bwire, H. 2011. Children's independent mobility and perceptions of outdoor environments in Dar Es Salaam City, Tanzania. Global Studies of Childhood, 1(3): 185–206 (Special Issue).

Bwire, H. and Chacha, P. 2011. An assessment of factors affecting the independent mobility of children in Dar es Salaam, 30th Southern African Transport Conference: Africa on the Move, Pretoria.

Collins, D.C.A. and Kearns, R.A. 2005. Geographies of inequality: child pedestrian injury and walking school buses in Auckland, New Zealand. *Social Science and Medicine*, 60: 61–69.

Department of Transport (DoT). 2005. *National household travel survey 2003: Technical report*. Pretoria: Department of Transport.

Engwicht, D. 1993. *Reclaiming our cities and towns: Better living with less traffic*. Philadelphia: New Society Publishers.

Fredericks, K. 2003. *Improvement of child and learner travel conditions in Cape Town: with specific reference to the planning of a hypothetical walking school bus*. Final year Civil Engineering undergraduate thesis, University of Cape Town.

Hillman, M. 2006. Children's rights and adults' wrongs. *Children's Geographies*, 4(1): 61–67.

Kearns, R.A. and Collins, D.C.A. 2003. Crossing roads, crossing boundaries: empowerment and participation in a child pedestrian safety initiative. *Space and Polity*, 7(2): 193–212.

Kingham, S. and Ussher, S. 2005. Ticket to a sustainable future: an evaluation of the long term durability of the Walking School Bus programme in Christchurch, New Zealand. *Transport Policy*, 12: 314–323.

Mackett, R.L., Lucas, L., Paskins, J. and Turbin, J. 2003. A methodology for evaluating walking buses as an instrument of Urban Transport Policy. *Transport Policy*, 10: 179–186.

Mackett, R.L., Lucas, L., Paskins, J. and Turbin, J. 2005. *Walking buses in Hertfordshire: lessons and impacts*. London: Centre for Transport Studies, University College London.

Muchaka, P., Behrens, R. and Abrahams, S. 2011. Learner travel behaviour and parent attitudes towards the use of non-motorised modes: Findings of school travel surveys in Cape Town. Paper presented at the 30th South African Transport Conference, 11–14 July, Pretoria.

O'Fallon, C. 2007. *Developing school-based cycle trains in New Zealand*. Land Transport New Zealand, Research Report 338.

11 The use of microscopic simulation modelling techniques to assess and predict road safety through an analysis of road user and infrastructure interaction in Cape Town

Rahul Jobanputra

Introduction

During the past decade the challenges posed by the number of road traffic crashes has gained prominence as one of the developing world's most pressing health and development concerns (WHO, 2009). In comparison to any international standard, the road safety record in South Africa, especially in relation to the most vulnerable group – pedestrians – is appalling, at around 15,000 fatalities per year (RTMC, 2010). This record has continued unabated over the last decade, and with the increasing rates of urbanization and motorization being experienced countywide, it is likely that fatalities and injuries, as a result of road traffic crashes, will continue to be a major issue unless greater emphasis is placed on road safety.

The underlying reasons for the fatality rates are complex. They are influenced by a combination of road network planning and design, the settlement patterns, and by behavioural and law enforcement issues. Of particular relevance to this investigation is that the road network planning and design concepts promote a car-based infrastructure system, comprising of a hierarchical system of provision of high-speed arterial and distributor roads, and limited, or no, provision for non-motorized travel outside of the central city areas. A compounding effect is the historic settlement patterns in South African cities, which force the urban poor to walk or use public transport, mostly along arterial routes with the concomitant danger of road crashes.

Recent legislative and policy environment for town planning, transport and road infrastructure provision is firmly aimed at the provision of improved public transport and NMT access. However, despite their obvious safety ramifications, it is likely that benefits will only accrue slowly.

In the meantime, the historical approach of addressing road safety – reactive assessments of historic crash data to determine so-called hazardous locations – are still used by local authorities and practitioners, despite the knowledge that road crash data is unreliable and does not contain sufficient detail for a comprehensive investigation.

Evidence from countries with low road fatality rates shows that a focused and dedicated systems approach to road safety yields significant benefits.

Further, the literature details many examples of complementary evaluation methods and predictive modelling to proactively assess or predict safety at particular facilities. However, as the majority of these techniques focus on vehicles it remains a concern that, despite carefully organized infrastructure and behavioural rules, shared surfaces, inevitably, lead to some kind of conflict with non-motorized users suffering the most.

Through case studies in Cape Town, and some innovative use of micro-simulation modelling, this chapter shows that modelling can be used to provide a better understanding of the interaction between the road user and the infrastructure to evaluate the potential benefits of engineering counter-measures and to provide a comparative evaluation of the likely safety characteristics of urban road infrastructure, with different operational characteristics, without the need, or use, of historic crash data.

Background

Travel in Cape Town is dominated by walking and public transport (DoT, 2013), and urban travel patterns are predominantly influenced by apartheid-era spatial planning, which segregated city centres/economic opportunities from the labour force by means of green buffer spaces and controlled transport corridors.

Partly as a consequence of this planning ideology, and partly because of the adoption of the pre-70s US style of road planning, roads in South Africa are based on a hierarchical system with little in the way of pedestrian facilities outside of the main city areas which, paradoxically, are the areas with the highest population densities. Significantly, in terms of vehicle–pedestrian interaction, this (citywide) system results in levels of service that allow high vehicular speeds on arterials and distributors. In addition, to cater for the large expansion in the population and unplanned areas, more recent road infrastructure of a similar standard, with commensurate travel conditions and facilities, has been provided (Jobanputra, 2013).

Despite walking being the dominant mode of transport for the vast majority of people and walking trip lengths being considerably longer than those conventionally assumed in practice (around 800m) (Behrens, 2005), there is a dearth of non-motorized transport (NMT) facilities, which inevitably adds to the conflict with motorized transport. Furthermore, the inconsistent nature of infrastructure provision means that pedestrian routes, if they exist at all, are not interconnected systems that would allow safe, efficient passage coinciding with desire lines.

Consequently, and also because of behavioural issues, South African cities have one of the worst road safety records in the world, at around 30 fatalities per 100,000 population every year (see IRTAD, 2009; OECD, 2012; RTMC, 2010). This level of annual road fatalities has continued unabated over the last decade, and of particular concern is that with the rapid increase in urbanization and motorization being experienced in the country, the number of fatalities is likely to rise.

Data from the South African National Injury Mortality Surveillance System (NIMSS) shows the following characteristics for transport related deaths: a high percentage of male deaths (70-80%), a high percentage of pedestrian deaths (55-60%) and high alcohol-relatedness of deaths among both drivers and pedestrians (>50% of deaths); there are distinct peaks over weekends among adults and, in the mornings and early afternoons, among children of school going age (MRC, 2007). The level of pedestrian fatalities compares poorly with the OECD's 26 member countries – which range between 8% and 37% of all road fatalities (ITF, 2011). The NIMSS and the City of Cape Town also report that crashes involving pedestrians, resulting in both serious[1] and fatal injuries, form the largest proportion of the known crash types in Cape Town.

Gender and age data indicate that males in the 26- to 40-year-old age group appear to be most at risk from crashes, as pedestrians. A significant number of these fatalities occur on the peripheries of the city, mainly on arterials, which are poorly enforced in terms of speed compliance.

The NIMSS also highlights that behavioural factors that reduce road users' ability to act safely, such as the consumption of alcohol, substance abuse, aggressive driving, the use of cell phones and fatigue, have a large part to play. In addition, there is a diversity of culture in South Africa with different behaviours and perceptions, which prevail within the traffic mix and can be witnessed every day.

Transport policy and legislative environment

Immediate solutions to reducing the number of pedestrian fatalities that spring to mind, and seem fairly straightforward, are to reduce arterial speeds (by some kind of enforcement), reduce the number of conflict points by provision of greater NMT facilities and encourage a greater use of public transport. The policy environment has recognized this (see, for example, NDoT, 1996, 2009) and strategies to develop priorities and implement new facilities are being actively pursued at the local level. The government is also aware of the disproportionate road traffic death toll borne by pedestrians in cities.

In parallel, South Africa is a signatory to the goals agreed in the Global Plan for the Decade of Action for Road Safety 2011–2022 (the 'Moscow Declaration'), and many of the pillars of this plan are being followed by responsible provincial authorities.

Despite these efforts, road safety continues to be a persistent problem, particularly in urban areas. It could be that these policies and implementation programmes have yet to take effect, and that it is only a matter of time before their effects on road safety will be felt, albeit that changes will be limited over time due to budgetary constraints and there needs to be a balance with the provision of an acceptable level of service for private vehicles.

Road safety concepts, indicators and predictive modelling

The traditional method of addressing road safety issues is to investigate 'hazardous' locations through historic data or newsworthy incidents, based on or via an engineering lens. From this, causes are attributed to human error, vehicular factors or the road environment. The implication is, therefore, that there is consistent, longer term, accurate and comprehensive historic crash data. But, any review of the crash databases in the country shows that they clearly do not meet these requirements. Hence, unless there is an obvious engineering shortcoming in the infrastructure provided, any conclusions drawn from this data may be flawed. Furthermore, this method is not proactive and does not address the root cause – the vehicle–pedestrian–environment interaction.

Not surprisingly, within such a complex socio-technical system, human error is consistently implicated as the overwhelming causal factor attributed to road traffic incidents, through investigations using historic data.

Literature on human error as a cause of crashes points to many sources of error, such as level of experience, fatigue or inattention to the driving task, alcohol and so on. It also shows that humans are willing to take what is termed as an 'acceptable risk' and, for instance, speed excessively, in favour of one or more utility factors (for example, see Wilde, 1998; SWOV, 2012).

The problem of adopting an approach that tries to identify individual components of the road user system to interpret crash data is that it produces typologies that mix up disjointed phenomena. Consequently, a systems approach, which focuses on the relationship between the parts that connect them into a whole, is generally preferred and has been adopted by countries that have achieved significant road safety benefits over the last decade (see, for example, Sweden's 'Vision Zero').

A systemic approach assumes that to handle complex behaviour, fundamentally, a connection needs to be made between the three components of the system in a manner that defines the system, what it does, what it becomes and its goal (i.e. the safety of the Human–Vehicle–Environment System). Taken from this perspective, a crash is the result of an incorrectly adjusted interaction between the system's components. The cause of a crash should, therefore, not be seen as a problem in one component or the other, but in the defective inter-component interaction.

The level of detail required for a systems approach is typically not available [for example, pre-crash conditions, detailed driving data (i.e. data that may be available from in-car video recorders), pedestrian data and actual detailed crash causes]. As a result, researchers have framed analytic approaches to study the factors that affect the number of crashes occurring in geographical space (usually a roadway segment or intersection) over some specified time period (week, month, year or number of years). Such an approach handles the spatial and temporal elements associated with crashes, and ensures that adequate data is available for the estimation of statistical models (in terms of measurable explanatory variables).

The literature also shows many examples of different crash prediction models (see Lord and Mannering, 2010 for a comprehensive listing) and other statistical models. The use of naturalistic driving simulators to investigate driver behaviour has also evolved recently. But, while such studies provide unique data, there is a practical limitation to the number of vehicles that can take part, which limits data availability. Researchers are also using near-crashes in combination with recorded crash events to provide insight into prevailing risks.

From an NMT perspective, predictive models have been developed to specifically estimate road safety risk to pedestrians. A comprehensive review of published predictive models for pedestrian safety has been compiled by the NCHRP (2008). It indicates that the most common form of predictive model uses the negative binomial structure and adopted the following general functional form for its predictions:

$$Nped = \exp(\beta 0 + \beta 1 ADT + \beta 2 PedVol + \beta 3 X3 \dots \beta n Xn) \quad (11.1)$$

where $\beta 0 \dots \beta n$ are coefficients to be estimated. *Nped* is the expected number of pedestrian crashes, *ADT* is the annual average daily traffic, *PedVol* is the annual average daily pedestrian volume, and $X3 \dots Xn$ represent other site characteristics such as proportion of left-turn volume, number of lanes, speed limits, absence/presence of crosswalk and absence/presence of a median.

The NCHRP (2008) study concludes by developing a prediction methodology for vehicle–pedestrian collisions (but only at signalized intersections) based on the above-mentioned function with *accident modification factors* (AMF). The cited advantage of these predictive models is that they can be readily applied to conventional intersections with minimum data, but their primary weakness is the limitation of the availability of crash data to generate a good model that can explain observed variations.

Once again, the majority of these predictive models or statistical techniques require historical crash data. The questionable nature of historical data has led to a move away from its use and towards more proactive safety analyses and planning.

Typical of this trend has been the use of observable near-crash events and other surrogate safety indicators. The former has been prompted by the fact that near-crashes occur far more frequently than crashes but have similar underlying processes, which gives them key advantages for their use in safety analysis. Safety indicators are a more resource-efficient and ethically appealing alternative for effective safety assessment. Methods available can be used as a research tool to establish the link between behaviour and risk, and for safety diagnosis (i.e. the problem at a site or series of sites).

Surrogate or proximal safety indicators

Commonly accepted criteria for transport safety applications put forward by Tarko et al. (2009) are that surrogate measures should satisfy two basic conditions: '(1) A surrogate measure should be based on an observable non-crash event that is physically related in a predictable and reliable way to crashes, and (2) there exists a practical method for converting the non-crash events into corresponding crash frequency and/or severity.'

The first condition emphasizes the crucial aspects of crash surrogacy that enable meeting the second condition: the development of a method of converting the surrogate outcomes into the meaningful outcome – frequency and severity of crashes.

Traffic volume meets the first condition of a good surrogate measure – vehicles must be present on the road for crashes to happen. However, this measure has a limited use as it seldom meets the second condition because, with the exception of traffic calming measures, most safety treatments do not affect traffic volume.

Speed is also proposed and used by some authors as a safety surrogate measure. It is an important component of a surrogate event definition, but its use as a stand-alone surrogate measure may be difficult due to the complexity of the speed–safety relationship. Other surrogate measures of safety proposed include actual traffic conflicts, time critical events, post-encroachment time, time-integrated time-to-collision, headways, shockwaves and deceleration-to-safety time. Traffic conflicts are, by far, the most prevalent measure considered by highway safety engineers (Tarko et al., 2009).

Some, or all, of these surrogate safety indicators can be used to measure the spatial and/or temporal proximity of safety-critical events. They are assumed to have an established relationship with crashes, and have the advantage of being more frequent than crashes (near misses, for example) and, therefore, require a shorter period of study to establish statistically stable values. They are also responsive to the specific characteristics and conditions of particular traffic locations or facilities, making them useful in before-and-after study designs, and other safety assessment strategies. A critical requirement of proximal safety indicators is the establishment of their validity, i.e. how well they represent actual 'safety' as a theoretical concept. Other questions concern the reliability of the various measurement techniques, their advantages and disadvantages and their usefulness from a practical perspective.

Despite extensive development of many techniques which use surrogate safety indicators (see, for example, the Swedish Traffic Conflict Technique), many national transport research institutes have raised reliability and validity questions surrounding these techniques, as well as asserting that they are diagnostic instruments only (see for example, Chin and Quek, 1997; Cunto, 2008). These arguments emphasize the need for proximal safety indicators to be useful in their own right without the need to validate against the measure of crash occurrence.

In summary, it is apparent that each observational technique has advantages and disadvantages, and different resource requirements. They all require trained and experienced observers and/or specialized video equipment and analysis, which is in short supply in South Africa. Allied to this is the critical aspect of inter-observer difference and the need to focus on pedestrians, rather than vehicles, as they bear the brunt of road traffic incidents.

The combined aspects of these factors led to the investigation of the feasibility of the use of microscopic simulation as a tool to assist in road safety analysis/prediction.

The use of microscopic simulation models in road safety

Microscopic traffic simulation allows dynamic traffic modelling that can provide a flexible test environment to study road user performance effects of new and alternative designs, and allows other safety-influencing factors related to the roadway, such as speed and flow, to be estimated. Although the main focus of traffic simulation models is, to a large extent, to help predict transport infrastructure requirements and traffic efficiencies, recent advances mean that models can now simulate the interaction between vehicles and pedestrians and, therefore, the safety implications of infrastructure provision where space is shared. This can be at a street, precinct or even suburb level, both temporally and spatially, as simulations can be modified to suit most local conditions through local input data and adjustable parametres.

Microscopic traffic simulation provides a number of clear advantages over more traditional traffic analysis tools in that it can provide a comprehensive set of results for an entire study area. It allows real-time visualization that is often valuable as a preliminary form of validation (for expert and non-technical audiences), and it can be tested in a virtual environment before implementation – which is especially appealing in situations where budgets are limited and where geometric and operational changes would be expensive and, possibly, troublesome.

The assessment of safety through simulation can be carried out proactively without the need for crash data, as models can output predicted details of proximal safety characteristics such as speed, flow and headways, along with safety-relevant interactive processes, such as gap acceptance in yielding situations.

The formulation of viable safety evaluation methodology, or indicators, based on the outcomes of micro-simulation modelling of specific urban contexts, therefore, offers significant benefits. In addition, the simulation should allow a better understanding of the possible interaction between the infrastructure, vehicles and pedestrians and, because of this, it should facilitate a better evaluation of the road safety risk of the situation under consideration.

In theory, therefore, micro-simulation models have the potential to provide a useful platform for many different types of evaluative and predictive safety analysis, and represent an alternative to more traditional statistical modelling. The main potential of simulation-based methodologies is likely to be preliminary forms of analyses, in the early stages of research, development and design, as well as in the assessment of comparative before-and-after type scenario studies to establish the effects of safety enhancements, or safety influencing measures, in the road environment at specific locations, and in relation to specific road-user groups. Simulations can also be used to generate many types of data simultaneously (safety, traffic performance and capacity and environmental impacts). This allows the analyst to get a more complete and comprehensive picture of the many different operational effects related to a particular area of study.

However, a significantly higher level of modelling detail is required for safety assessment than for other traffic system objectives. This is particularly evident for pedestrian and behavioural sub-models that describe the interactive processes and provide the simulation output. For safety analysis, it is of critical importance to ensure the accurate representation of interactive behaviour between road-user entities and their interaction with the environment. It is also important to ensure that the parametres reflecting road-user behaviour and vehicle performance are appropriate to the local conditions. These higher levels of modelling fidelity require the collection of more detailed empirical data and demand greater stringency in the processes of model calibration and validation.

There are many existing commercial and academically developed micro-simulation programmes. Commercial products provide regular support and functional improvements as a result of ongoing research and development work that is necessary to meet clients' needs, market forces and the ever-evolving computing industry. Typical of these types of simulation packages are PARAMICS, VISSIM, AIMSUN and CORSIM. Each one has its own peculiarities and strengths and is usually supplied with additional functions and features with each new version.

In relation to modelling road-user behaviour for specific situations (such as overtaking on single lane carriageways, representative modelling of car-following and gap-acceptance behaviour, heterogeneous traffic and the evaluation of safety and environmental impacts), commercial software has limited functionality; however, user-programmable interfaces allow the possibility of tailoring models to suit situations that encompass most of the factors that have a direct or indirect influence on road safety.

These facets are confirmed in the literature (see, for example, Algers et al., 1997; Cunto, 2008; Young and Archer, 2010). However, it is evident that, to date, there have been few attempts to include the more vulnerable road users in any of these studies, or to link simulation modelling to the more straightforward safety surrogates of speed and volume by testing the interaction between road users on safety-related infrastructure measures like traffic calming devices. The above-mentioned software should, therefore, be able to provide a platform for the development of more holistic safety assessments that apply mechanistic microscopic approaches, rather than observational approaches, to evaluate the relative safety of infrastructure.

To determine whether this was the case, case studies were conducted in urban areas of Cape Town, where the majority of the safety-critical events in the city occur, particularly, pedestrian fatalities. The infrastructure at these locations is sufficiently different in form and function to present many user and behavioural differences. The context adopted was, therefore, broad enough to adequately capture the diversity of users and uses in the city, and to provide a reasonable test for the capability of the software. This chapter reports on two case studies. One case study deals with the evaluation of surrogate safety, as a result of changes to vehicular outputs for traffic calming measures, and one deals with the interaction between vehicles, pedestrians and the infrastructure.

Simulated safety performance measures are not required to correlate directly to the actual number of crashes, but the relative difference of various intersection designs, measured by simulated safety performance, must be consistent with similar studies of real-world conflict measurements. To test the utility of the simulation approach to road safety, three criteria for the simulation output were determined: (a) it should discriminate between the safety of two design alternatives; (b) there should be a proportional correlation between surrogate safety measure reductions and predicted reductions in traffic conflicts; and (c) it should correlate (in terms of a surrogate safety measure) with real-world traffic conflict levels.

Case study I: the assessment of the safety effects of road-based traffic calming measures by micro-simulation

The strong relationship between speed and crash potential, as well as crash severity, is well established – lower speeds allow more reaction time for all road users, as well as reduce the likelihood of serious pedestrian injury, especially at speeds below 50 km/h. This speed threshold, and lower, is widely used in urban areas where vehicles and pedestrians are likely to interact.

In South Africa, a study of vehicular speeds of vehicles, carried out by the RTMC (2005), found that around 30% of drivers exceeded the 120 km/h freeway speed limit, indicating that general levels of compliance to speed limits in South Africa are particularly low. Non-compliance of speed limits also tends to be exacerbated by other attitudinal, behavioural and road environmental factors. An example of the road environmental factor is shown in a study by the FHWA in the US (Stuster et al., 1998) which reports that drivers who had travelled at 70 mph (102 km/h) for more than three minutes tended to drive

5 to 15 mph (8-24 km/h) faster in a 30 mph (48 km/h) zone than drivers who had not previously driven at the faster speed.

Road-based traffic calming is fundamentally concerned with reducing the adverse impact of motor vehicles in urban areas; it usually involves reducing vehicular speeds and providing safer crossing points, or more space for pedestrians and cyclists, thereby improving the 'liveability' of the local environment. With increasing congestion, transport policies are attempting to shift transportation towards walking and cycling and to make environments more attractive for walking in terms of safety, efficiency and convenience, especially in more densely populated areas, where pedestrian–vehicular interactions can have a significant impact on each other.

Traffic calming measures consist of engineering and other measures (perceptual or psychological) that can be implemented to achieve a change in driver behaviour and, through them, a reduction in the adverse impact of motorized transport in urban areas. Such measures have been successfully implemented in cities across the world, for decades, in response to safety-related neighbourhood traffic concerns.

The range of calming measures in current use is fairly extensive. Types used vary depending on the application area and desired effect. Application areas can vary from minor changes to local streets to area-wide strategies, the latter being preferable because of the ability to assess and control the nature and effects of diversionary traffic.

As stated, micro-simulation modelling should provide a more scientific basis of assessment of the relative safety of potential measures. The method has the added benefits of applicability and uniformity of analysis: not only does it allow specific traffic facilities to be modelled dynamically in great detail, in a safe offline environment, but it can represent a large number of factors that have a direct, or indirect, influence on traffic safety and performance that can be varied in a model, including flow rates, turning movements, average speeds and speed variance, signalling, various aspects of road-user behaviour, and aspects related to site geometry and design.

However, the usefulness of this methodology (in terms of safety assessment) relies on its ability to capture the overall level of turbulence for different transportation scenarios as a function of a number of attributes. The results must be able to provide meaningful insights about changes in overall safety for different engineering counter-measures. For example, what would the safety benefits (or dis-benefits) be of speed humps compared to chicanes? Are the results sensitive to changes in spacing, volume and different modes of transport? And, by extension (with the use of the safety surrogates of speed and volume), what would the relative reductions in fatalities be for each measure modelled?

As a representative of other packages, the PARAMICS software suite was used to model a road network approximately 1.5 km long and 7.3 m wide near to the centre of the City of Cape Town. The network formed part of a model encompassing the outer edges of the City, which allowed the possibility of diversion routes being used by vehicles to avoid any increases in their generalized cost of travel due to the interventions imposed. This provision allowed both volume and speed effects to be measured for each intervention considered from the simulation outputs.

In addition to normal geometric and functional data for the entire area, empirical data was collected from a range of traffic calming devices from sites around Cape Town to inform behavioural values for interventions considered and to calibrate the model. This included driving speeds, which were collected using an in-vehicle GPS device and a cell-phone application, and headway and deceleration. Headway was determined using general empirical data from the area as a whole (to ensure consistency in approach and validity of the simulation), whereas deceleration was determined from measured values and based on literature related to the average age of the vehicle.

As micro-simulation models are generally aimed at modelling and assessing network performance and optimization, the modelling of traffic calming measures is seldom straightforward – the software is not designed to incorporate short and sudden changes in either vertical or horizontal alignments (which are the prevalent features of most traffic calming devices). User-defined modifications and appropriate manipulations to provide a representative outcome are the usual workarounds in such instances. Despite this possibility, models are limited in their ability to simulate the effects of devices, such as rumble strips or different types of textured paving. Modelled interventions for this case study were, therefore, limited to speed humps, mini-roundabouts, chicanes, raised intersections, a narrowing and a 'road diet'.

A statistical comparison between all the traffic calming measures modelled and the base case is provided in Table 11.1. As expected, mean speed values are less than the base case and vary between measures. The simulated mean speed value over speed humps was predicted to be the lowest (although this is a function of its size and spacing). The standard deviation value for the measures is an indication of the dispersion of simulated speeds around their mean value. It can also be assumed to represent the degree of safety of the measure because large variations in speed along a stretch of road can present difficulties for both drivers and pedestrians. Furthermore, the variation in speed from the mean can also be a proxy for the environmental impact of the measure as it is an indicator of the level of acceleration and deceleration levels, which relate directly to emissions. The conclusion from this analysis is that a road 'diet', which does not induce large variations in speed, would be the best calming measure, especially if it could be implemented in a way that forces a greater speed reduction.

It can also be concluded that this software can be used to optimize the design of new traffic calming applications – from the type of applications to be used to achieve the desired outcome, to a combination of applications and their spacing. The decision-making process using this type of analysis would, thus, be more scientifically grounded and, in addition, it could encompass the effect on public transport options.

Table 11.1 Statistical comparison of simulated traffic calming measures

Statistic	No calming	Speed humps	Mini-roundabout	Chicanes	Raised inter-section	Narrowing	Road diet
Arithmetic mean speed (km/h)	49.7	32.7	43.1	38.2	41.2	43.8	42.3
Maximum speed (km/h)	54.3	48.2	51.6	49.2	51.7	53.5	43.7
Minimum speed (km/h)	38.7	21.7	29.8	25.8	24.2	28.9	38.3
Standard deviation	4.3	6.2	5.9	7.3	8.2	6.7	1.3

A review of contemporary literature on common international road-based traffic calming strategies and approaches was undertaken to compare published speed and volume reduction values with the results of the simulated outputs for the same applications (and, thus, to validate the simulation method used). From the comparisons, it was clear that there was a large degree of commonality; however, values in relation to the case of speed humps were significantly different to published data. It is postulated that this is not as a result of a flaw in the approach used, because of possible differences in (a) the before and after speed limits; (b) the point at which speeds were measured; (c) the time of day of the assessments; (d) the physical design of the measure(s); and (e) generalized driving behaviour and differences in law enforcement/driver compliance levels. All of these aspects have an influence on the level of effectiveness of a particular measure and, given that the simulation output matched local empirical evidence, it is representative of local conditions.

The simulated results of traffic calming measures on traffic flow, for the range of measures considered, on the other hand, were more consistent with published results.

The overall conclusion that can be drawn from this work is that the simulation method used will yield results for speed and traffic flow effects that would be reliable and that would provide details of varying effects, depending on the type of traffic calming measure tested. The method can be used for single measures, a combination of measures and even area-wide strategies; and the level of impact predicted would be realistic for each case.

This confirms the first objective of the utility test for the software – that the model can produce results that differentiate potential speed and traffic flow changes for different types of infrastructure interventions. The results also show that the model is sensitive to different transport modes, indicating that it could be used to analyze the potential safety impacts of different transport strategies, such as public transport or freight corridors.

Tests of the model, in relation to changes in spacing of measures, show that its output (for speed, volume and travel time) is also sensitive to changes in spacing, as would be the case in a real-life scenario.

The second objective of this analysis was to assess the level of impact that was likely from a particular intervention and, from this, the corresponding change in safety risk and crash severity (i.e. fatal, injury, no-injury) and that the simulation outputs, specifically speed and volume, can be used as surrogates for safety risk. Their impact on safety has already been confirmed by, amongst others, the WHO (2004) – generally, volume reductions provide adequate time and gaps to allow safer pedestrian crossings, cyclists to move in greater safety, and a general reduction in levels of exposure. Speed reductions, on the other hand, are associated with a reduction in the risk of crash occurrence and its severity.

In terms of the level of risk reduction as a result of reductions in vehicular speeds and volumes, the literature indicates that estimated safety benefits vary between each study. For instance, the estimated benefits of a 25% speed reduction in fatal injuries vary from a 35% reduction (using Elvik, 2009) to 83% (using NCHRP, 2008). However, these findings confirm that the greater the speed reduction, the greater the likely benefit. The evaluation of potential safety impacts, as a result of changes in traffic volumes due to traffic calming, is much more complex and case-specific. This is because (i) changes in volumes due to calming measures usually affect the entire network, not just the street in consideration; (ii) changes in volume are affected by the availability of alternate routes; (iii) the application of other measures in area-wide treatments may have as large an impact on volumes as the geometrics and spacing of calming measures applied to the street in question; and (iv) more significantly, volume impacts depend, fundamentally, on the split between local and through traffic, even though studies conducted (see, for example, Ewing, 1999) confirm that traffic calming measures will not affect the amount of locally bound traffic, unless they are so severe or restrictive that they 'degenerate' motor vehicle trips.

Despite these complexities, studies (see, for example, Eenink et al., 2008) show that a reduction in volume does provide a commensurate decrease in fatality levels.

Case study II: assessment of road safety implications due to road user infrastructure interaction by micro-simulation

The work in case study I confirms that micro-simulation models can successfully simulate the potential safety effects for a range of traffic calming measures using traffic speed and volumes as surrogates for safety. This case study adds to it by assessing the potential of simulation models to estimate the relative levels of safety for road infrastructure in different locations and provision level and with different road user characteristics, within the Cape Town metropolitan area.

The two areas selected for this study were Lansdowne Road in the southeast part of Cape Town (Study area A) and Coen Stytler/Buitengracht Street (Study area B) in the city centre (see Figures 11.1 and 11.2).

Over the last 8–10 years, the City's reported crash statistics indicate that both areas have been the scene of many collisions, with a large proportion of pedestrian fatalities or injuries. In particular, Lansdowne Road regularly features in the annual list of worst known locations for road traffic fatalities. The majority of the fatalities were reported as having occurred during the peak periods, with the highest numbers being during winter.[2]

Although it is not possible to replicate actual crash numbers by means of simulation, the aim of the study was to establish whether the simulation model can replicate the (un) safety of the infrastructure and to consider whether the output allows a distinction to be made between the two areas, in terms of their relative levels of safety.

The Lansdowne Road corridor is approximately 10 km in length (total). It is a major urban arterial linking 'low-cost' residential areas (known as 'townships') to the Cape Town CBD and other employment areas/activity nodes. The adjacent land uses and direct (informal and unplanned) access to the corridor can be seen to create major traffic conflicts and flow issues.

In contrast the Coen Stytler/Buitengracht Street intersection provides access to two of the major areas of employment and interest in Cape Town – the CBD and the Victoria and Alfred Waterfront (V&A)[3] (Figure 11.2). The intersection is located on the fringes of the CBD and is flanked on the southern side by hotels and on the northern side by landscaped areas and a

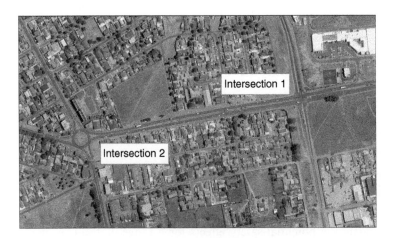

Figure 11.1 Study area A: Lansdowne Road and vicinity.
Source: City of Cape Town, data received 2016.

multi-story car park. It consists of multi-lane approaches and exits and is heavily trafficked throughout the day because it provides access to and from the two main national routes in the Western Cape, the N1 and N2. The intersection is controlled by a four-stage signal system and all approaches and exits are subject to a 60 km/h speed limit.

The southern side of the intersection (Approach B) is a major crossing point for workers travelling on foot from the railway station (in the southeast) to the dockyards (to the north) and back, as well as for many tourists travelling to and from the V&A Waterfront. This level of pedestrian activity and vehicular traffic resulted in 206 serious pedestrian crashes and 27 fatalities in the period 2000-2008 (City of Cape Town, data received 2010). The vast majority of the fatalities occurred at the site of the main pedestrian desire lines – at the intersection and at some distance away from it – presumably because of the lack of appropriate crossing facilities, and despite a pedestrian signal phase at the intersection. The City's crash records indicate that most of these incidents occurred during the morning and evening peaks when pedestrian flows are at their highest. Notably, and presumably in response, the City constructed a pedestrian overbridge over half the road (i.e. only over the exit of Buitengracht!) in 2010.

Along with the normal/key data required for model building and calibration, vehicle and pedestrian speeds, travel times and, where applicable, queue lengths were collected. Vehicle speeds and travel times were again collected from a combination of a GPS-enabled smart phone application and the in-vehicle GPS based vehicle tracking system. Pedestrian data – numbers, walking speeds and compliance levels/risk-taking behaviour – were captured manually through field observations and by use of video recordings. Other major

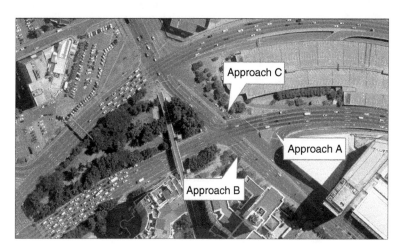

Figure 11.2 Study area B: the Coen Styetler/Buitengracht Street
           intersection.

Source: City of Cape Town, data received 2016.

pedestrian parametres in the PARAMICS suite required assessment through sensitivity tests and comparison to observed flows.

From a review of the literature of the surrogates for road safety, it was felt that an appropriate and reliable indicator for safety in this study would be to use deceleration rate. Values and safety definitions were based on proposals adopted by Hydén (1996) but modified to account for modern vehicle efficiencies and systems.

The simulation output provides a raw-data file containing deceleration and trajectory details for each vehicle from which it is possible to identify temporal and spatial positions of each vehicle. Safety-critical events were identified as those where vehicles exceed a 7 m/s braking threshold. A simplified classification of the more common forms of incidents that could occur, at either of these study areas, is indicated in Figure 11.3, and is used in the analysis for both study areas. All other types of conflicts are not considered, owing to their low frequencies and relevance to this study.

From these files it was clear that the majority of incidents in both areas A and B were related to vehicle–pedestrian conflicts. To identify vehicle–pedestrian incidents in more detail, a feature of the software – an 'agent collisions' viewer – is used. As implied, this is a graphics file which provides a visual recording of potential vehicle–pedestrian incidents and their locations. A data file logging these events is also generated by the simulation and this, in combination with the vehicle trajectory file, details the time and place of

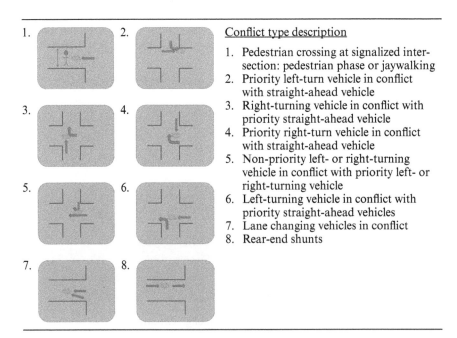

Conflict type description

1. Pedestrian crossing at signalized intersection: pedestrian phase or jaywalking
2. Priority left-turn vehicle in conflict with straight-ahead vehicle
3. Right-turning vehicle in conflict with priority straight-ahead vehicle
4. Priority right-turn vehicle in conflict with straight-ahead vehicle
5. Non-priority left- or right-turning vehicle in conflict with priority left- or right-turning vehicle
6. Left-turning vehicle in conflict with priority straight-ahead vehicles
7. Lane changing vehicles in conflict
8. Rear-end shunts

Figure 11.3 Classification of potential conflict types

predicted incidents and provides the requisite details for a thorough review of likely pedestrian incidents.

The number of collisions generated were, however, far too large for the simulation period (in excess of 60) and, thus, a modified approach called a Potential Collision Index, based on vehicular speed at the point of 'impact', was adopted (see Table 11.2).

Using this index, Study area A was assessed as having a PCI of 54 for a one-hour simulation period.

As an additional consideration, and as an extension to case study I, appropriate infrastructural modifications (mainly in the form of traffic calming measures) were investigated to establish their effect on predicted vehicle–pedestrian incidents and, through these, to demonstrate relative benefits likely from the changes.

Using the same methodology, a PCI of 29 was obtained for a modification involving the incorporation of raised crosswalks at appropriate locations – i.e. a 55% increase in safety – and, based on the trajectory files, speeds related to predicted incidents were below the threshold that fatalities are likely to occur (based on Rosen and Sander's study of 2011).

Similarly, the results for a changed configuration to allow bus-only routes on both sides of the road indicated that there would be no change from the base case in terms of the PCI. This was due, not to the number of incidents, which were forecast to decrease, but to the severity – it was forecast that bus lanes would allow some vehicles to speed excessively. This outcome raises the possibility that there may be issues with the pedestrian gap acceptance model and visibility envelope specifications (particularly the vertical plane, in terms of buses).

Again, using the same methodology for study area B (Coen Stytler intersection), an analysis of the severity of braking events (see Figure 11.4) indicated that only two events could be categorized as being potential 'collisions' and eight others where vehicle speeds exceeded 20 km/h. These results contrasted

Table 11.2 Proposed Potential Collision Index (PCI)

Rating	Criteria/description
0	Collision unlikely (vehicle speed <20 km/h)
1	Pedestrian recognized – swerve or stop likely (vehicle speed <30 km/h and decelerating)
2	Collision possible with minor consequences (speed 20–30 km/h, no deceleration)
3	Collision possible with medium consequences (speed 30–40 km/h)
4	Collision possible with serious consequences (speed 40–60 km/h)
5	Collision likely with serious consequences, chance of fatality >50% (speed 60 km/h+)

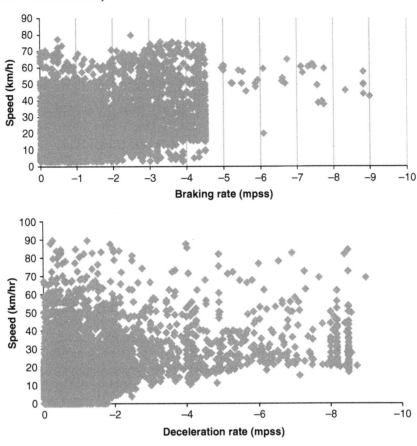

Figure 11.4 Case study II, area B: severity of braking events during simulation
period

with a modification to the signal timing to allow an additional 7 seconds of
total pedestrian crossing time and a change in the surfacing texture of the exit
showed that, in the former, there was likely to be a 50% reduction in the number of 'collisions' and, but for a change in the surfacing texture, there may be
little or no change in safety in terms of numbers of incidents, but that if they
occurred, the severity would be decreased.

The conclusion from these tests is that the application of micro-simulation
using similar (and probably more refined) methods would have the benefits
of providing a useful and scientific method of evaluation of road safety, both
proactively and retrospectively, as the method fulfils the utility criteria prescribed. Further, simulation tests can be carried out on any type of infrastructure in a comparative format, studies without any historic data, in an
off-line environment without the need for implementation and without undue
investigator differences. The results from such studies can be used by planners or decision-makers to prioritize possible improvements or as part of a
road safety campaign. And, although studies can be carried out remotely, i.e.

anywhere (and by anyone with the requisite modelling skills) the case studies showed that the local content is paramount in terms of road-user behaviour and that, despite taking due reasonable care in numerical calibration, inconsistent behaviour can be viewed through the animation output.

The studies did reveal several software issues that require addressing and/or user programmable interfacing if this methodology is to gain widespread acceptance. Most of these relate to the non-compliant nature of road users in South Africa (and probably the other parts of Africa), particularly, the interaction between jaywalking and vehicles could not be fully captured, as the pedestrian behaviour algorithm does not allow for irregular stops/starts/rapid acceleration/deceleration, yielding between lanes and a re-calculation of gap acceptance during crossing. In addition, the consideration and further development in automated post-processing systems for simulation outputs would help alleviate the requirements for manual evaluations of their outcomes, and enable an evaluation of longer simulation periods which would ultimately provide better and more averaged assessments of safety performance.

Notes

1 A serious injury is defined as one where a person has sustained injuries to such an extent that hospitalization is required (www.arrivealive.co.za).
2 The winter period in Cape Town usually runs from June to August and is characterized by periods of fairly heavy rainfall. December is a period of exceptional numbers of fatalities due to an annual migration of workers from dormitory townships surrounding most large conurbations to their 'ancestral' homes or places of birth.
3 The Victoria & Alfred Waterfront in the historic heart of Cape Town's working harbour is South Africa's most-visited destination, having the highest rate of foreign tourists of any attraction in the country (www.sovereign-publications.com/waterfront.htm).

References

Algers, S., Bernauer, E., Boero, M., Breheret, L., Taranto, C. di, Dougherty, M., Fox, K. and Gabard, J. 1997. Review of micro-simulation models. Review Report of the SMARTEST Project; deliverable 3, 1997. Available at: www.its.leeds.ac.uk/smartest/deliv3.html.
Behrens, R., 2005. Accommodating walking as a travel mode in South African cities: Towards improved neighbourhood movement network design practices. *Planning Practice and Research* 20, 163–182.
Chin, H.-C. and Quek, S.-T. 1997. Measurement of traffic conflicts. *Safety Science* 26, 169–185.
Cunto, F. 2008. *Assessing safety performance of transportation systems using microscopic simulation*. Ontario: University of Waterloo.
DoT. 2013. National Household Travel Survey 2013. Statistical release P0320. Pretoria: Statistics South Africa.
Eenink, R., Reurings, M. (SWOV), Elvik, R. (TOI), Cardoso, J., Wichert, S. (LNEC) and Stefan, S. (KfV), 2008. Accident prediction models and road safety impact assessment: recommendations for using these tools. RIPCORD-ISEREST, Sixth Framework Programme. European Union.

Elvik, R. 2009. The power model of the relationship between speed and road safety. Update and new analysis, TOI report 1034/2009. Oslo: Institute of Transport Economics, Norwegian Centre for Transport Research.

Ewing, R., 1999. *Traffic calming: state of the practice. Publication no. IR-098.* Washington, D.C.: Institute of Transport Engineers.

Hydén, C. 1987. The development of a method for traffic safety evaluation: the Swedish traffic conflict technique. Department of Technology and Society, Lund University, Lund, Sweden.

Hydén, C. 1996. Traffic conflicts technique: State-of-the-art. In: Topp H.H. (Ed.), 1996. *Traffic safety work with video-processing. Green Series No.43.* University Kaiserslautern. Kaiserslauten, Germany: Transportation Department.

International Transport Forum (ITF). 2012. *Sharing road safety: Developing an international framework for crash modification functions.* Paris: OECD Publishing. DOI: http://dx.doi.org/10.1787/9789282103760-en

IRTAD. 2009. International Traffic Safety Data and Analysis Group. Accessed at: http://internationaltransportforum.org/irtadpublic/index.html.

ITF. 2011. International Transport Forum, Road Safety Annual Report, 2011. Accessed at: www.internationaltransportforum.org/pub/pdf/11IrtadReport.pdf.

Jobanputra, R. 2013. *An assessment of the applicability of microscopic simulation models to assess and predict road safety for vehicle-pedestrian and infrastructure interaction in developing cities.* Thesis submitted for PHD, Civil Engineering, Centre for Transport Studies University of Cape Town, 2014.

Lord, D. and Mannering, F. 2010. The statistical analysis of crash-frequency data: a review and assessment of methodological alternatives. *Transportation Research Part A: Policy and Practice*, 44, 291–305.

MRC. 2007. A profile of fatal injuries in South Africa. 7th Annual Report of the National Injury Mortality Surveillance System, Section 4. Cape Town Metropolitan Area. Cape Town: Medical Research Council - University of South Africa.

NCHRP. 2008. Pedestrian safety prediction methodology, Project 17–26. Washington, DC: National Cooperative Highway Research Programme, Transportation Research Board.

NDoT. 1996. White paper on national transport policy. Pretoria: National Department of Transport.

NDoT. 2009. National Land Transport Act. Pretoria: National Department of South Africa.

Rosen, E. and Sander, U. 2011. Literature review of pedestrian fatality risk as a function of car impact speed. *Accident Analysis & Prevention*, 43: 25–33.

RTMC. 2005. *Speed as a contributory factor to road traffic crashes.* Pretoria: Road Traffic Management Corporation.

RTMC. 2010. Road traffic report 31 December 2010. Pretoria Road Traffic Management Corporation.

Stuster, J., Coffman, Z. and Warren, D. 1998. Synthesis of safety research related to speed and speed management. Publication no. FHWA-RD-98-154. Federal Highway Administration. Accessed at: www.tfhrc.gov/safety/speed/spdtoc.htm

SWOV. 2012. The relation between speed and crashes. SWOV Factsheet. Liedschendam, The Netherlands: SWOV Institute for Road Safety Research.

Tarko, A., Davis, G., Saunier, N., Sayed, T. and Washington, S. 2009. White Paper. Surrogate Measures of Safety ANB20(3). Subcommitee on Surrogate Measures of Safety, Safety Data Evaluation and Analysis – ANB20. Washington, DC: Transportation Research Board.

WHO. 2004. World report on road traffic injury prevention. Summary. Geneva: World Health Organisation.

WHO. 2009. Global status report on road safety: time for action. Geneva: World Health Organisation,.

Wilde, G.J., 1998. Risk homeostasis theory: an overview. *Injury Prevention: Journal of the International Society for Child and Adolescent Injury Prevention*, 4: 89–91.

Young, W. and Archer, J. 2010. The measurement and modelling of proximal safety measures. *Proceedings of the ICE - Transport* 163: 191–201.

12 Institutional framework for walking and cycling provision in Cape Town, Dar es Salaam and Nairobi

Winnie V. Mitullah and Romanus Opiyo

Introduction

Many African cities have not given walking and cycling modes the importance they deserve. Walking accounts for over 45 percent of the modal split in many cities across Africa (see Chapter 2), but this prominence is not reflected in the provision of infrastructure and related facilities. In order to contribute to the understanding of this situation, this chapter examines actors, their relationships and interactions, and how they influence the delivery of walking and cycling solutions in Cape Town, Dar es Salaam and Nairobi.

Methods

The chapter is based on a review of primary and secondary sources of information, which focused largely on the theoretical basis of actors in transport, the status of current policy on walking and cycling in the three cities, identifying different actors and their interaction or influence in shaping walking and cycling policy in the three cities, and practical efforts at implementing or improving institutional coordination on walking and cycling policy.

Actors in urban transport policy

The role of actors is essentially a question of agency in policy formulation and implementation. Transport policy change does not just happen on its own. Regulations, policies, knowledge and practices are institutional agencies mobilized to create solutions through different actors who require coordination and synergy (Dyrness, 2001). Policy intervention and change is not a one-off event but rather an iterative learning process that may involve success and failure at different periods of time (UN-Habitat, 2013). There are several actors and relationships involved in urban transport policy development and implementation. The key ones are government agencies, the real estate sector, public transport providers, urban transport planners and civil society (Vasconcellos, 2001).

Government agencies

An important issue to analyze is the capability and role of the state in service delivery, including different forms of provisions (public, private and partnership), subsidies and embedded debates in relation to public transport, walking and cycling. This kind of analysis requires an analysis of the nature, duties and responsibilities of the state with regard to transport and the extent to which the state makes an effort to fulfil the commitments it has to the citizens, as spelled out in its *social contract* with the public in the form of election manifestos, constitution, transport policies, short- and long-term plans, and strategies. Transport infrastructure has been provided, mostly by the state, and both public transport services and users have received legal or financial support during critical phases, when adequate provision or affordable access was threatened. This central role of the state continues to be important even when considering the recent neoliberal policies to deregulate or privatize public services and infrastructure provision in Africa.

Real estate sector

One of the most powerful stakeholders is the real estate industry, in view of the financial capital it controls. Its power may be exerted directly on the city authorities or on the members of the city council. Such influence is often directed to change or adapt the legal environment surrounding the right to acquire and use land. The real estate sector may implement its projects while ignoring the needs of walkers and cyclists.

Public transport providers

The provision and spatial distribution of public transport services greatly affects walking time and average conditions related to the quality of sidewalks. Public transport may be provided by public agencies or by private operators. In the case of private operation, it may be provided by individuals with their own vehicles, by cooperatives of bus owners or by bus enterprises. When bus enterprises have long-term contracts they acquire a disguised monopoly over the area or the lines they operate. Such conditions may make it extremely difficult for public authorities to change the supply pattern. Every proposal to change that appears to threaten the operators' profits or dominance is fiercely opposed. When individual operators form cooperatives, they become even stronger in the protection of their business.

Urban transport planners

A number of transport engineers in the developing countries were trained as 'highway' engineers, under the technical realm in the tradition of the US and Europe. Their daily practice has generally focused on providing road space for vehicles, especially automobiles owned by elites. They seldom care and plan for the movement of walkers, cyclists and public transport vehicles. Virtually all cities in Africa, and many other countries in the world, have detailed maps of the road systems but rarely maps of the sidewalk system or bicycle routes. The sidewalks are not considered a public task and rely on what local residents may do to build and maintain them. In view of their key role in technical elements in design, and their use of statistical techniques, most highway engineers prioritize 'highway' development options that will, supposedly, solve the mobility problems of all people (Monheim, 1996).

Such an ideological approach is partly supported by bias and prejudice against walking and cycling within the elite and by the ideology of motorization as a symbol of progress, regardless of the social costs that have to be paid to achieve it. Though developing country transport professionals (planners, engineers, economists, geographers, sociologists etc.), and development agencies and banks (Hook, 1995), have, until recently, tended to accept the assumptions and methods used in high-income countries, which, as discussed previously, tended to favour motorized over walking and cycling travel, there is a shift and professionals are advancing the cause for the unique and important roles that walking and cycling play in an efficient and equitable transportation system in all cities. It is, therefore, important to analyze the practical experiences, especially the various obstacles professionals face from political groups (affluent voters, businesses and special interest groups, such as taxi owners) that support automobile-oriented planning. Some of these professionals have made an effort to redress the long-established planning bias that favours motorized travel in Africa, and elsewhere. Their efforts provide insights into how politics, economics and interest groups can hinder or promote walking and cycling policy, and shows the extent to which the new transport planning tools (Litman, 2013) are being used to change the practice in different contexts.

Civil society

This category of actors have not previously been active in the public transport sector, but this is beginning to change, as reflected in a growing number of civil society organizations and associations in Cape Town (such as Bicycle Cape Town, the Cape Town Bicycle Map and Open Streets Cape Town) and in Kenya, where Two Wheels Power, a local organization which advocates for cycling in Kenya, and the Kenya Alliance of Residents Association (KARA) are becoming active in influencing policy and practices and creating awareness among NMT users and providers, by sensitizing both users and infrastructure providers. These are initiatives that require scaling up for popularization of walking and cycling. Embedded in the relationships and processes, among the different actors, are the institutions which they create and, in turn, influence their behaviour and conduct.

Embedding walking and cycling in transport policy in Cape Town, Dar es Salaam and Nairobi

Research conducted by ACET revealed both progress and challenges facing walking and cycling in Nairobi (Mitullah and Opiyo, 2012), Dar es Salaam (I-ce, 2000) and Cape Town (Republic of South Africa, 2009). In Nairobi, the few facilities provided are in different stages with others located along motorized roads without bollards (see Chapter 7). Bollards have been fixed to protect property, including fuel pumps, lighting poles and traffic lights and, in some cases, as an after-thought after losing lives, bollards have been fixed to provide safety to pedestrians and cyclists. Jogoo Road, during the initial improvement, provided an example of lack of integrated planning for NMT (see Box 12.1), and continual attempts to improve the situation.

A review of transport policy, in respect to walking and cycling in the three cities, reveals that planning and provision for walking and cycling, in Nairobi and Dar es Salaam, is largely externally driven by development partners; a strong proactive strategy by the cities is only beginning to emerge. Cape Town has a more rooted planning practice for walking and cycling than the other two cities.

There have been several iterations in the development of walking and cycling policy in these three cities and, at present, there is a policy document on NMT in each city. A summary of the NMT policy content follows.

The City Council of Nairobi (CCN) Strategic Plan for 2006-2012 did not provide for NMT, although it indicated that the road network would be enhanced. However, with support from UNEP, the Nairobi City County Government prepared a draft NMT policy in March 2015. The policy aims to

Box 12.1

When the ACET study began in 2009, Jogoo Road corridor in Nairobi had no NMT facilities and conflict between NMT and motorized transport was serious, with many deaths. During the research process, the road underwent a renewal with NMT infrastructure provided along the road without any bollards to protect NMT users. This did not help the NMT users, and during peak hours the NMT facility became a parallel highway for motorized vehicles. Stones and boulders at several points to block the motorized transport from accessing the facility provided interim measures to protect NMT users. This partially resolved the problem, but wheelchair users had to detour, while pedestrians either detoured, climbed or jumped over the stone barriers, thereby defeating the purpose of NMT infrastructure provision. To address these ad hoc approaches, the authorities, eventually, did what should have been done in the first stage of planning – fixed bollards along the corridor, resulting in walking becoming safe with many people using the NMT facility.

develop and maintain a transport system that fully integrates NMT as part of the Nairobi transport system. The policy further provides for:

- Creation of a safe, cohesive and comfortable network of footpaths, cycling lanes and tracks, green areas and other support amenities
- Laws and regulations to ensure prioritization of NMT facilities
- Promotion of investment in walking and cycling infrastructure
- Mass rapid systems to be connected with walking and cycling facilities and an accessible city centre with improved safety and reduced energy use and emissions

For Dar es Salaam, the NMT issues are embedded in the National Transport Policy (NTP), which was prepared in the year 2003. The vision of the policy is: 'To have efficient and cost-effective and international transport service to all segments of the population and sectors of the national economy, with maximum safety and minimum environmental degradation' (United Republic of Tanzania, 2003). With reference to NMT, the policy provides for the following:

- Design of residential areas should ensure security, safety and comfort- ability to pedestrians and cyclists by providing for dedicated pedestrian and cyclist lanes.
- Design should also consider planting of flora, including trees and flow- ers along the urban roads, to provide for both attractive road scenes and shading to pedestrians from overhead sun.
- Influencing land use planning and resettlement patterns to achieve easy access to amenities.
- Road reserves should not be used for purposes that hinder smooth flow of traffic and future expansions.

The NMT policy for Cape Town has an overall vision of creating an enabling environment in order to become a prosperous city that achieves effective and equitable service delivery. The transport goals are set in line with the National Transport Policy, which focuses on achieving a significant modal shift from private to public transport, and extending the NMT network. The objectives of the NMT policy include (City of Cape Town, 2005):

- Increasing cycling and enabling walking as modes of travel
- Creating a safe pedestrian and cycling environment
- Developing a quality, attractive and dignified environment
- Promoting a changed culture that accepts the use of cycling and walking as accepted means to move around in the city and elicits more responsible behaviour

The NMT network in Cape Town is receiving attention, but proper links and NMT provision are still lacking in most areas. The policy identifies challenges, such as lack of NMT alternatives for short distance travel (Jennings, 2015).

Under the South African Provincial Integrated Transport Steering Group, the establishment of land-use incentives and NMT improvements of around 10 under-developed public transport nodes, of provincial significance, were to be addressed by 2014 (Provincial Key Projects) by ensuring that:

- Dedicated NMT Expanded Public Works Program projects are in place by 2014.
- Every provincial road project in the province must include a NMT component.
- NMT Plans must be developed and implemented for each municipality of the Province, as a part of the mobility strategy and IPTN rollout by 2014.
- Dedicated cycle lanes in the Western Cape must be doubled by 2014.
- Links should be added to logically extend the catchment area of public transport stations and improve the inclusiveness of existing neighbourhoods.

There are a number of issues, relating to NMT, which cities need to address through planning, provision and protection of walking and cycling facilities. While developing a policy is an important step in planning, its translation into routine practice, and provision of infrastructure in the work programmes and plans, is where true commitment to the policy is shown. Consequently, it is necessary to assess whether the different actors are taking the necessary steps, not only to formulate policies, but also to implement the policies.

Actors in walking and cycling policy implementation

The urban transport sector attracts many interests and actors, including citizens who require efficient modes of transport; government agencies charged with the responsibility of policy, planning and implementation of programmes; and investors keen on investing in public transport. While motorized modes of transport have attracted attention of many actors and stakeholders, the NMT sector has been less attractive. This is in spite of the fact that NMT is used by almost all city residents and addresses not only health issues, but also environmental issues of pollution and congestion (Kinney et al., 2011).

The ongoing retrofitting of NMT infrastructure, in the cities of Nairobi and Dar es Salaam, is not being driven by the city authorities but by the development partners in collaboration with the central governments. The entry of development partners, in particular the World Bank, African Development Bank and the Japanese International Cooperation Agency (JICA), was triggered by the African Union's initiative of a programme on Transport and Millennium Development Goals. In most of the cities in Africa, the AU initiative and the World Bank supported the Sub-Saharan Africa Transport Programme (SSATP) and related studies influenced the promotion of NMT provision. This has seen several countries initiating the development of integrated national transport policies, with few cities cascading the policies to city levels. However, implementing public transport programmes remain a major investment, which cities on their own cannot handle. In response to this challenge, the Republic of Kenya has established the Kenya Urban Roads Authority (KURA) to deal with urban roads, while the United Republic of Tanzania has yet to come up with an urban transport policy and a specialized authority.

In Nairobi, the mandate of KURA includes NMT infrastructure and planning. Most of the NMT facilities in Nairobi are provided along the roads which KURA oversees. KURA is mandated to advise both the Ministry of Transport and the city authorities on how best to plan and ensure ample space for the provision of NMT infrastructure and the protection of NMT from encroachment by motorized vehicles. While the example of Jogoo Road, provided above, was initiated before the establishment of KURA, it is questionable why incremental development of NMT infrastructure continues without protecting NMT users, even after the establishment of KURA.

Although the city authorities of Nairobi do not have grand programmes of NMT provision, city authorities began by formulating by-laws which protect NMT users, including provision for observing traffic lights and zebra crossings (www.nairobi.go.ke/home/common-city-laws-and-regulations/by-laws). The by-laws provide for NMT as a legal mode of transport requiring protection, but the main challenge is the enforcement of the by-laws. Enforcement has also been ad hoc and poor. With the support of UNEP, the city formulated a draft NMT policy which provides direction for NMT. This policy is expected to complement the integrated planning provided in the Urban Areas and Cities Act, 2012 for effective planning and integration of NMT provision in city infrastructure programmes.

The Kenya Police, through the Traffic Department, plays a key role in management of traffic within the city but does not seem to care about NMT users. This manifests in traffic police control of vehicles at traffic light intersections without giving adequate time to NMT users to cross. The situation is worse in busy intersections where there is a constant spill-over of vehicles, with the traffic police unable to maintain a coordinated flow, giving each joining road/lane a slot. In such intersections, pedestrians constantly mingle with motor vehicles as they make individual attempts to cross the intersections, while traffic police call on motorized vehicles to drive through.

Apart from the traffic police, other government ministries, departments and agencies, such as the Ministry of Transport, Ministry of Health and Kenya Roads Board, also deal with different aspects of transport. The Ministry of Transport drove the formulation of Integrated National Transport Policy and worked with a number of partners in undertaking studies and financing development of infrastructure. However, the policy formulation lagged behind the implementation of programmes, which were largely undertaken in an incoherent manner.

As noted earlier, the provision of NMT facilities is largely influenced by the development partners, who are also being influenced by local, regional and international advocacy on integration of NMT in public transport. The agencies have committed financial and technical resources in commissioning studies and funding several projects in Kenya and Tanzania. A number of these studies, especially those by the JICA, World Bank, SIDA and the African Development Bank, have contributed significantly to the visibility of NMT infrastructure in many urban road projects and to the development of policy and the National Road Safety Action Plan, 2006/2010 for Kenya. The plan covered many aspects and addressed the safety needs of NMT, including provision of infrastructure and enhancing national emergency capacity to deal with victims of road accidents.

In Kenya, the development partners have supported interventions on NMT facilitated by policy directions on NMT in the Traffic Act, Cap 403 and agreements between development partners and the government. The Traffic Act provides for all county governments to develop by-laws for managing traffic, while agreements between the government and development partners have enabled the agencies to support the city urban mobility projects. This has been done through modest investments in NMT facilities in new road construction and along major arterial roads, and pedestrian paths. Such NMT is envisaged to link the areas where the urban poor live, with locations where they walk to work (World Bank, 2006). The pro-poor interventions have, mainly, been undertaken through the Kenya Municipal Programme (KMP), covering capacity building of city government in areas of policy development and institutional and financial management relating to design, implementation and management of higher level infrastructure.

The private sector also plays a crucial role in public transport. However, in the area of NMT, the local private sector is more of a hindrance, rather than a solution. The paratransit mode, which is largely operated by private entrepreneurs, has been a major threat to NMT users, and a cause of accidents and deaths. Most paratransit modes, *matatus* in Kenya, *dala dalas* in Tanzania and *taxis* in Cape Town, encroach on NMT facilities, especially during peak hours, causing serious conflict, often resulting in accidents and deaths. In Kenya, although the entrepreneurs belong to three different associations covering owners, welfare and drivers and conductors, they do not focus on any aspect of NMT, including safety, which they constantly threaten. There has also been limited interest of local entrepreneurs in investing in NMT. The same case applies to the non-state actors who have not been active in creating awareness and educating citizens on NMT.

Mitullah and Opiyo (2012) have observed that the NMT agenda is being driven, mainly, by development partners who fund roads. The partners have made provision for NMT funding conditional for road projects, albeit with minimal coordination. The pressure from partners has created partial awareness among relevant government ministries, related agencies, such as the Kenya Roads Board, KURA and urban authorities. This has resulted in

NMT being recognized as a standard feature of any road project in Nairobi, although the major challenge remains lack of standard NMT design, which results in inappropriate retrofitting of footpaths.

The NMT infrastructure across the city of Nairobi is not standard, due to lack of a national standard manual guiding the design of NMT provision and its enforcement. Existing infrastructure relies on design skills of individual engineers carrying out road works. This results in many NMT facilities, such as footbridges, speed bumps and pedestrian crossings, not being utilized, due to poor design and inappropriate location (Mitullah and Opiyo, 2012). Overall, existing retrofitted NMT facilities are not integrated into the public transport system within the city. The facilities are implemented in an ad hoc manner, and in isolation of other modes, in response to larger global concern of improving active transport by promoting NMT. This leaves NMT users de-linked from other modes and land-use activities, which they engage with on a daily basis.

The case of Tanzania is not different from Kenya. The country has many stakeholders involved in public transport but the agencies hardly know what each other is doing, especially in the area of urban development. Kanyama and Goran (2009) note the weaknesses of formal institutional coordination in planning for public transport in Dar es Salaam and Nairobi. This may be attributed to the fact that none of the existing Acts of Parliament and legal provisions, mentioned in the previous section, deal with urban transport, nor do they provide clarity over which institution has overall responsibility for urban transport planning and management, let alone NMT provision. In theory, for Kenya, the Ministry of Transport (MoT) should provide policy and strategic direction and planning for the sector, including that of NMT. However, the lack of a national urban NMT policy, ineffective urban planning and the many uncoordinated actors result in the inefficiency of the urban transport system. The urban authorities have neither the power nor the resources required for addressing the complexity of responsibilities in urban public transport. The Dar es Salaam Agency for Rapid Transit (DART), which is largely overseeing the development of Bus Rapid Transit (BRT), is also responsible for orderly traffic flow, redesigning intersections, and improving traffic management. The Dar es Salaam BRT will incorporate facilities for walking and cycling, which will encourage use of NMT (Takule, 2010).

South Africa has a number of stakeholders involved in walking and cycling policy, with all three levels of government having a role in transport. The responsibility for designing and implementing NMT, largely, lies with the public sector. Involvement of non-governmental organizations (NGOs) and the private sector is purely voluntary, but consultation with the two sectors is paramount for planning for required facilities. The National Department of Transport is the public agency responsible for policy and legislation for issues of national concern. The department also provides guidance to provinces and municipalities, such as the city of Cape Town, as well as the development of guidelines and regulations.

The provincial governments in South Africa assist district municipalities and coordinate and approve the Municipal Integrated Transport Plans (MITP) and business plans for NMT funding. The city of Cape Town seems quite proactive in integrating NMT, as highlighted in its NMT policy provisions. The city also sets budget aside for the NMT network and universal access, as well as pedestrianization. In 2014/2015 and 2016/17 a total of R80,000,000, for each of the years, was allocated to NMT (Jennings, 2015).

Public transport providers, consumers and the private sector, including experts, also play a role. The private sector, through provision of subsidy and Cooperate Social Responsibility (CSR), have a role in supporting NMT and encouraging employees to use NMT and providing relevant facilities, including shower and bicycle parking facilities (Republic of South Africa, 2009).

Collaboration and coordination among actors

In almost all African countries, changes to centralized planning have been taking place since the beginning of the 1990s, with many public organizations embracing New Public Management (NPM), which requires participation of all stakeholders. Prior to this, planning in most cities in Africa relied on statist centralized approaches to planning and management. This resulted in the exclusion of most stakeholders, largely due to lack of policies, strategies and plans for attracting stakeholders into activities of the state. Without policy direction from governments and well-outlined strategies, plans, and national standards guiding the design of NMT provision, it becomes quite difficult to attract other stakeholders. Authors, such as Wekwete (1997), have noted that the dominance of central authority in development paradigm hampers involvement of other actors and undermines local autonomy.

Stone (1990) appreciates the usefulness of informal and unplanned linkages among organizations, as opposed to large central authority; while Miller and Lam (2003) argue that formal forms of coordination are better for coordinating operations that involve numerous stakeholders. Although coordinating public and private organizations and institutions remain a challenge, institutional analysis cannot ignore any of the sectors. What is relevant is synergy between the two, since a range of factors, from personal attributes to organizational structures, contributes to the successful coordination. This notwithstanding, there can never be efficient coordination without policy direction, strategies and plans. It is these provisions that provide opportunity for stakeholders to meaningfully participate in any development undertaking.

This NPM has brought new actors, themes and new sites of engagement. This shift from conventional political institutions to organizations actively negotiating between sovereign bodies and inter-organizational networks, that challenge the established distinction between public and private sector (Hajer and Wagenaar, 2003), has begun improving coordination in areas with clear policies, strategies and plans. UN-HABITAT pioneered promotion of the stakeholder approach through their Sustainable Cities Programme, which promotes and coordinates participatory management of urban development. The programme identified urban mobility as a fourth issue in urban governance after environment, health and sanitation; ecological planning and open space management; and institutional challenges (UN-HABITAT, 2001).

In Kenya, the participation of stakeholders in development processes is embedded in the Kenya Constitution promulgated in August 2010, and related legislations. The provision requires citizens, through their organizations and associations, to participate in all aspects of development, including policy development, planning and management of projects. The County Governments Act, 2012 obligates county governments to promote public participation, including that of non-state actors [Section 104(4)]. It, further, gives county planning units the responsibility of coordinating integrated development planning within respective counties, and ensures meaningful engagement of citizens in planning [Section (105(5)]. These provisions, in addition to ongoing reforms, are expected to change the manner of planning within cities, including that of transport and related NMT provision.

Coordination of actors has been a major concern in the transport sector, due to the many actors involved in the industry (Asingo and Mitullah, 2007; Kanyama and Goran, 2009; Government of Kenya, 2012) and the high expectation for coordination from analysts. Most analysts begin from the assumption that coordination is good, without taking into consideration the fact that the many actors are guided by different visions, mandates, interests and regulatory regimes. Factors noted to constrain prospects for institutional coordination in planning for public transport include lack of city visions, effective city and public transport plans, professionalism and regulatory frameworks; rampant corruption, poverty and poor citizen and stakeholder participation; inadequate political and fiscal decentralization; and unwillingness of decision makers to change existing transport systems (Kanyama and Goran, 2009; Asingo and Mitullah, 2007).

The factors constraining participation should be seen as the institutional fabric within which organizations, managing public transport, operate. The cognitive-cultural, regulative and normative understanding of the public transport environment should be used in designing a working mechanism for collaboration and coordination. Often, this is not the case. Many analysts call for a common understanding of all the factors by all actors, a task which seems unrealistic within a complex institutional framework. What is required is a well-established coordination mechanism where actors exchange ideas, plan interventions in line with their strengths, and monitor and evaluate progress for the purpose of moving forward. Such a framework requires a minimalist approach with clarity of vision, policy and legislative framework of operation, most of which are lacking in African cities.

The actors engaged in NMT planning and provision have their mandates, and NMT provision is not included among their core mandates. In stakeholder mapping and analysis, the Global Environment Facility (GEF) proposal on Promoting Sustainable Transport Solutions in East African Cities notes that organizations, authorities and interest groups in the cities of Nairobi, Addis Ababa and Kampala have different motives and interests with regards to Mass Rapid Transport (MRT), Bus Rapid Transport (BRT) and NMT (UNEP, 2010). The case is not different for NMT in the case study cities, although the GEF proposal does not list NMT in this matter.

The lack of focus on NMT is partly due to lack of policy on NMT, and the dominance of motorized transport as a mode of public transport. In the case of Kenya, the formulation of a National Integrated Transport Policy has provided opportunity for actors, such as the Ministry of Transport, to enact relevant legislation, including legislation on NMT, to enable the implementation of policy and coordination of actors. This provided a good environment for cities, such as Nairobi, to develop relevant laws, including by-laws for improving NMT operations within the city and, in 2015, the development of a draft NMT policy. The same case applies to Tanzania, which should be able to use the Urban Transport Policy and the Dar es Salaam Transport Master Plan (JICA, 2008), prepared with the assistance of JICA, to develop guidelines and strategies for NMT.

Involvement and consultation, among actors, in planning and managing public transport is a recent experience. A study by Kanyama and Goran (2009) indicates a lack of consultation and involvement with relevant organizations, such as the Ministry of Health and the health departments of city authorities. Such departments play key roles, including handling victims of road crashes and regulating health aspects in vehicles, for example, ventilation, overcrowding of passengers, and awareness creation regarding health aspects of public transportation. Although the study attributes this to the emphasis on physical planning and infrastructure provision, dominated by central governments and urban authorities, the lack of institutional mechanisms for collaboration and coordination of actors is a major drawback to ensuring provision of an integrated public transport system.

The reality, in the cities of Nairobi and Dar es Salaam, reveals that ad hoc convenient mechanisms are often put in place when there are specific issues to be addressed. For example, in Kenya, the development of the Road Safety Action Plan (Government of Kenya, 2006), the Integrated National Transport Policy (2012) and facilitation of various international support programmes have used this approach. While this approach to planning for public transport, including NMT, provides an alternative approach for encouraging cooperation across agencies, such approaches do not ensure institutionalization of processes. The processes are not routine and are not informed by clear guidelines, standards and strategies aimed at promoting integrated and sustainable public transport. In the case of Cape Town, where the city has a framework which includes a strategy, as well as a Bicycle Master Plan and a Pedestrian Implementation Plan, proper links and NMT provision is still lacking in most areas (City of Cape Town, 2013).

However, projects such as Shova Kulula and walking buses, informed by embedded strategies and plans, are likely to survive political regimes as opposed to ad hoc projects previously implemented in the cities of Nairobi and Dar es Salaam.

The ongoing political reforms in Kenya have resulted in the establishment of a number of agencies with specialized mandates in public transport, with the respective offices charged with coordination and supervision for ensuring cohesion among government agencies. Responsibilities of such offices included improving service delivery to citizens and ensuring effective engagement with citizens and stakeholders. This coordination of the public transport sector, which is handled by many government agencies, has been a tall order. A collaboration of UNEP and KURA, through the Ministry of Transport, has produced a showcase of NMT for Nairobi under the Share the Road global initiative. The collaboration mobilized government resources and rehabilitated a major residential road leading to the UN complex. The facility has sidewalks on either side of the road, and a separated and protected cycling lane on one side of the road (www.unep.org/transport/shareroad/).

Coordination has been a major issue in Nairobi, especially when the Ministry of Nairobi Metropolitan Development was established and the Ministries of Roads and Transport split in two. This resulted in one ministry charged with regulation of road transport services and services and the other with infrastructure planning and programming. The splitting of public transport between two ministries was political and problematic for collaboration and coordination. However, the new governance system seals the number of government ministries to not more than 24, as opposed to the previous 42 ministries. The new system also provides a new governance framework for urban areas and cities, and gives prominence to participation of citizens through citizen forums. The legal requirements for developing an integrated development plan includes the provision of basic guidelines for land use management in consultation with stakeholders and citizen participation. The City County of Nairobi has developed an integrated development plan and protocols for operationalizing the plan are incrementally being developed.

In Nairobi, there is a proposal to create a Nairobi Metropolitan Transport Authority responsible for service planning, service management, quality control, control and sanction penalties and approval of fares. In this approach, national government ministries would retain policy and regulatory oversight over the authority. The former Ministry of Nairobi Metropolitan Development had the mandate of ensuring preparation and enforcement of an integrated spatial growth and development strategy and actualization of integrated strategic programmes for the provision of social, economic and infrastructure services (GoK, 2010), including NMT. The Ministry of Transport is also spearheading the Nairobi Metropolitan Region Bus Rapid Transit Programme (NBRTP). These planning initiatives do not mention NMT, and there is a need for advocacy on NMT as a policy, planning and implementation issue.

Policies provide direction, while regulations and guidelines order how programmes and plans are implemented and managed. In the absence of these important elements, activities are undertaken without any proper basis, and are liable to discontinuity. This challenge is made worse by the poor enforcement of existing regulatory provisions, bias towards motorized transport and poor coordination of public transport actors, among other issues. Policy and related legislation is useful for planning and management of any sector, but Kenya and Tanzania, for many years, implemented NMT projects with neither an NMT policy nor integrated public transport policy. This left urban development without any frame of reference, relying on political decisions of urban governors influenced by political interests and development partners (Mitullah and Opiyo, 2012). This is expected to change as cities develop NMT policies, as committed to during the Africa Sustainable Forum (ASTF) 1st Ministerial and Experts Conference of October 2014. The Ministers of Environment and Transport committed 'to integrate sustainable transport into the region's development and planning processes and increase the amount of funding going to sustainable transport programs in Africa' (ASTF, 2014).

Reliance on urban governors and development partners, without local policies, strategies and plans, is quite problematic due to the high turnover of urban governors and the inability of development partners to sustain programmes. Although development partners, such as the UN-HABITAT and UNEP, are advocating for integrated development through the Sustainable Cities Programme, embedding programmes, within the national vision and development agenda, remain the preserve of national governments and urban authorities. The cities of Nairobi and Dar es Salaam have lagged behind in developing policies, legislation and standards for ensuring an integrated national development policy, including NMT provision. In the midst of this process, ad hoc improvement and provision of NMT infrastructure has been going on, albeit without standards and guidelines. The Nairobi draft NMT policy is a positive move and it is expected that cities like Dar es Salaam will also develop NMT policies for effective planning and management of sustainable transport.

The minimal attention and minimalist approach given to NMT affects institutionalization, since the cities of Nairobi and Dar es Salaam have no dedicated budget for NMT, although the Nairobi draft NMT policy indicates that the city will dedicate 20% of all road budgets to NMT. Other ministries and road agencies combine the budget for motorized carriageway and NMT facilities, and it is not easy to isolate the budget dedicated to NMT. The Share the Road initiative in Nairobi promotes NMT by advocating increased investment by donors and governments in walking and cycling road infrastructure. The initiative calls for at least 10% funding dedicated to NMT for the benefit of all road users. However, this has not been possible to assess, in the case of Kenya, since road cost is not separated to enable identification of what goes into NMT (UNEP/Climate XL, Africa, 2009).

Conclusion

Analysis in this chapter reveals that, while there are several actors who can implement existing policies, their coordination and strategic movement towards implementing a transformative NMT policy remain a hindrance. The progress made by the city of Cape Town and the ongoing measures in the cities of Nairobi and Dar es Salaam for integrating NMT in governance are progressive, although there is still much more to be done. Progress made in Cape Town, symbolized by a number of NMT projects, shows that with operational policies, strategies, plans and standards, cities can address NMT needs and make cities more liveable and sustainable.

Besides the legal foundation for Kenya, the ongoing advocacy for sustainable transport, which gives prominence to walking and cycling as modes of transport, the existence of a first showcase NMT project in Nairobi, and ongoing inclusion of NMT on new roads and retrofitting of NMT infrastructure on existing roads, provides evidence of progress. The Nairobi government and the proposed Nairobi Metropolitan Transport Authority (NMTA) are expected to support ongoing programmes for NMT. The county government is also expected to begin operationalizing NMT policy, plans and strategies, and to fast-track the establishment of the NMTA, as well as come up with an effective mechanism for NMT provision and coordination of actors.

The ongoing NMT interventions are expected to continue with the support of governments. Embracing NPM to bring on board other stakeholders with varied resources, and capacity, bridges the policy gap. This type of synergy and intervention is not new in Africa. A number of cities have applied what Tanzania, during the late 1960s and 1970s, used in the implementation of Ujamaa Policy – the 'we-must-run-while-others-walk' approach. Goran Hyden refers to it as a problem solving approach, as opposed to reliance on policy. Hyden argues that, due to a sense of urgency, effective resolution of ends and means conflicts is never done prior to making 'big' decisions (Hyden, 1979). While this debate is long dead, this chapter reveals that the two case study cities of Nairobi and Dar es Salaam, to a large extent, are still using the same approaches in the provision of NMT; while in Cape Town, policies, strategies, plans and standards for NMT have existed for a while, and the city seems to be way ahead of the other cities in Africa.

Institutionalization of NMT in urban transport, in the three case study cities, requires a shift from bias towards motorized transport to an institutionalized integrated approach with NMT policies, strategies and plans. The shift should influence stakeholders to prioritize NMT in mobilizing resources, budgeting, funding and managing public transport. This will avail direction for NMT provision, as opposed to ad hoc measures, which are not well grounded with the budget assigned to NMT. Achieving this goal requires synergy among actors and mobilization of resources, including relevant stakeholders, since NMT infrastructure is costly and can only be effectively provided through a public/private partnership approach.

References

Africa Sustainable Transport Forum (ASTF). 2014. 1st Inter-Ministerial and Experts Conference. Nairobi: Gigiri, 28 September–3 October.

Asingo, P. and Mitullah, W. 2007. Implementing roads transport safety measures in Kenya: policy issues and challenges. Nairobi: University of Nairobi, Institute for Development Studies, Working Paper No. 545.

City of Cape Town. 2013. Comprehensive integrated transport plan (2013/2018). Cape Town: City of Cape Town.

City of Cape Town. 2005. City of Cape Town NMT policy and strategy. Cape Town: City of Cape Town.

Dyrness, G. 2001. *Policy on the streets: a handbook for the establishment of sidewalk-vending programs*. Center for Religion and Civic Culture. University of Southern California.

Government of Kenya (GoK). 2010. Development of a Spatial Planning Concept for Nairobi Metropolitan Region Draft Plan. Ministry of Nairobi Metropolitan Development, Nairobi.

Government of Kenya. 2006. National Road Safety Action Plan. Nairobi: Ministry of Transport.

Government of Kenya. 2012. Integrated National Transport Policy. Nairobi: Ministry of Transport.

Hajer, M. and Wagenaar, H. (eds). 2003. *Deliberative policy analysis: understanding governance in network society*. Cambridge: Cambridge University Press

Hook, W. 1995. Economic importance of NMT. Transportation Research Record No. 1487. Washington, D.C.: TRB, National Research Council.

Hyden, G. 1979. We must run while others walk: policy making for socialist development in the Tanzania-type of polities. Economic Research Bureau Paper No. 75.1. University of Dar es Salaam.

Interface for Cycling Expertise (I-ce). 2000. *The significance of non-motorized transport for developing countries: sharing strategies for policy development*. Ultrecht.

JICA. 2008. Dar-es-Salaam Transport Policy and System Development Master Plan. Dar-es-Salaam: JICA.

JICA. 2013. Master plan for urban transport in Nairobi Metropolitan Area. Nairobi: JICA.

Jennings, G. 2015. Finding our balance: considering the opportunities for public bicycle systems in Cape Town, South Africa. *Research in Transportation Business & Management*, 15: 6–14.

Kanyama, A.A and Goran, C. 2009. *In search of a framework for institutional coordination in the planning for public transportation in Sub-Saharan African Cities: an analysis based on experiences from Dar-es-Salaam and Nairobi*. Stockholm: Royal Institute of Technology (KTH), Department of Planning and Environment.

Kinney, P. et al. 2011. Traffic impacts on PM2.5 Air Quality in Nairobi, Kenya. *Environmental Science and Policy*, 14(4): 369–78.

Litman, T. 2013. *Transportation affordability: evaluation and improvement strategies*. Victoria Transport Policy Institute (VTPI).

Miller, M and Lam, A. 2003. *Institutional aspects of multi-agency transit operations*. Institute of Transport Studies. Berkeley: University of California.

Mitullah, W.V. and Opiyo, R. 2012. Mainstreaming non-motorized transport (NMT) in policy and planning in Nairobi: institutional issues and challenges. Paper presented at the South African Transport Conference (SATC), 2012. Pretoria: SATC.

Monheim, H. 1996. Mobility without cars. *Transport and Environment*, 21: 31–34.

Republic of South Africa. 2009. *Non-motorized transport in Western Cape draft strategy*. Department of Transport and Public Works, Provincial Government Western Cape.

Republic of South Africa. 2012. *City of Cape Town – Comprehensive Integrated Transport Plan 2006 to 2011*. The City of Cape Town's Transport Authority.

Stone, D. 1990. Organisation and transit performance in Bay Area: a theoretical and empirical review. Institute of Transportation Studies. Berkeley: University of California.

Takule, C. 2010. Development of urban transport infrastructure: the place and role of non-motorized transport (NMT). Dar-es Salaam City Experience. Presentation at the UNEP Share the Road Event – Nairobi, Kenya. 29 November 2010. Retrieved from: http://staging.unep.org/transport/sharetheroad/PDF/STR_PolicyBrief_final.pdf

UNEP - Global Environment Facility (GEF). 2010. Proposal on promoting sustainable transport solutions for East African cities. Nairobi: UNEP.

UNEP/Climate XL, Africa. 2009. Share the Road: minimum standards for safe sustainable, accessible transport infrastructure in Nairobi – final draft. Nairobi: UNEP.

UN-Habitat. 2001. Cities in a globalising world. Global Report on Human Settlements 2001. London and Sterling, VA: United Nations Centre for Human Settlements (Habitat).

UN-Habitat. 2013. The Tool for the Rapid Assessment of Urban Mobility in Cities with Data Scarcity (TRAM).

United Republic of Tanzania. 2003. National Transport Policy. Ministry of Communications and Transport.

Vasconcellos, E.A. 2001. *Urban transport, environment and equity: the case for developing countries*. London: Earthscan Publications.

Wekwete, K. 1997. Urban management: 'the recent experience'. In Rakodi, C (ed), *Urban challenge in Africa: growth and management of its large cities*. New York: United Nations Press.

World Bank. 2006. Kenya Municipal Program (KMP), Project Information Document.

13 When bicycle lanes are not enough

Growing mode share in Cape Town, South Africa: an analysis of policy and practice

Gail Jennings, Brett Petzer and Ezra Goldman

Introduction

Cape Town can with relative confidence claim its title 'Cycle City' – coined by the then Executive Mayor Dan Plato in 2011 when launching Africa's longest continuous bike lane[1] – at least when compared to other urban centres on the continent. Routinely cited in the official and media discourse as being the most bicycle-friendly city in Africa,[2] Cape Town has 415 km or so of marked facilities (other than public roads) on which bicycles are permitted (in many cases refurbished or reallocated sidewalks shared with pedestrians). The city also has a number of high-quality bicycle parking racks, makes provision for transporting bicycles on the bus rapid transit (BRT) vehicles,[3] and a dedicated NMT (non-motorized transport[4]) unit at Transport for Cape Town (TCT),[5] the local transport authority. But bicycle transit mode share has remained largely unchanged in 10 years (see also Chapter 2, and Jennings, 2015) and as low as 0.5% of daily trips (Transport for Cape Town, 2015). Cyclists bear a disproportionate risk of being killed or injured as a result of road traffic incidents or crashes (Jobanputra, 2013). The commitment of TCT to supporting cyclists and the quality of bicycle-friendly interventions is questioned by an increasingly vocal civil society:

> Some days I really don't feel like being diplomatic … about how sh_t it is for … cyclists across the city – and how 5 or 6 years later in some parts little to nothing has changed. Performance and impact remains low. Some cute 'add-ons' and 'goodenoughism', 'so thankful to haves' is certainly far from the shift people need…"
>
> (Facebook comment, 12 July 2015)[6]

The City's Transport for Cape Town (TCT), under-resourced by global standards, is grappling with a multitude of unmet mobility needs and significant transport disadvantage. Its bicycle interventions are limited by the challenges of retrofitting infrastructure within an existing built environment, among others, and a policy direction constrained by the national, provincial and local government emphasis on bicycle infrastructure as key to providing for current users and growing mode share.

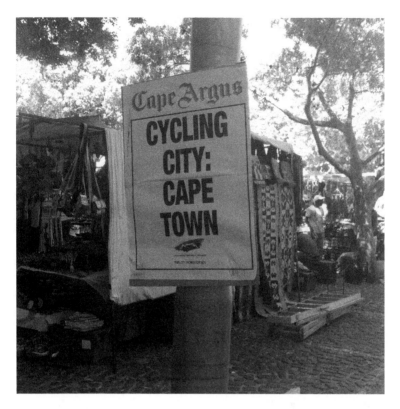

Figure 13.1 Newspaper poster, central Cape Town, 12 March 2012: Cape Town is
        Cycle City.

Policies and programmes to increase the rate of bicycle transport (or utility
cycling) in South Africa place great emphasis on the need to create physical
bicycling infrastructure. Although the policies have referenced the need for
what we refer to in this chapter as 'soft' infrastructure – the promotion of a
'culture and respect of NMT' (Transport for Cape Town, 2015; City of Cape
Town, 2005), supportive law enforcement, vehicle-free days, and financial
incentives such as showers and bicycle travel allowances – the actual outcome
of such policies so far has been the creation of under-used, relatively disconnected and inconsistent hard infrastructure.

This chapter will first provide an overview of South Africa's national and
local 'infrastructure-first' approach, and will argue that this narrow approach
evident in Cape Town until recently may go some way to explaining the lack
of significant mode shift in the city since the first walking and cycling policy.
We then consider the conditions beyond infrastructure that have given rise
to a bicycle culture and increased mode share in cities elsewhere – Bogotá
(Colombia) and Copenhagen (Denmark), in particular. In order to do so,

we draw on a variety of sources, including personal interviews, public policy, official city communication, city reports and social media.

Theoretical framework

Cities globally face challenges when attempting to increase utility cycling (Handy et al., 2014). Justifying the allocation of resources and road space is almost always a hard sell.

Our literature review revealed that what little published work there is on bicycle transport in South African (and African) cities focuses on NMT infrastructure or facilities guidelines and their quality, or on analysis of the inadequate implementation of NMT policy (e.g., Vanderschuren et al., 2015; Labuschagne and Ribbens, 2014; Mashiri et al., 2013; Bechstein, 2010; Gwala, 2007; Ribbens and Gamoo, 2006). Chapter 8 assesses the design, provision and quality of such infrastructure for both utility cycling and walking in Cape Town.

In their international review of infrastructure, programmes and policies to increase bicycling, John Pucher et al. (2010) note that studies evaluating the impact of infrastructure do not 'adequately address the direction of causality, such as whether bicycling infrastructure led to increased levels of bicycling, or whether bicycling demand led to investments in bicycle infrastructure'. Based on their earlier work in the US, Krizek et al. (2009) had reached a similar conclusion – that facilities might be the effect, rather than the cause, of high bicycle use because the people lobbied for the construction of such facilities. Work conducted in the United Kingdom (Goodman et al., 2013) finds that while infrastructure is well used, it primarily attracts current cyclists.

Emblematic cycling success stories, such as those of Bogotá (Colombia) and Copenhagen (Denmark) which influence Cape Town's aspirations, exhibit or have increased cycling's modal share through targeted investment in both 'hard' (infrastructural) and 'soft' (non-infrastructural) measures. This distinction between 'hard' and 'soft' infrastructure is drawn from McClintock (2002a), and is intended simply to contrast engineering-led approaches to understanding cycling, which are prevalent in low-cycling contexts (where they often constitute the entirety of official planning for cycling), with approaches drawing on the social sciences, in which aspects such as culture, attitudes, behaviours, discourses and beliefs are given greater weight in cycling planning.

The 'soft' approach, the aim of which has been defined by Cox (2013: 1) as "an attempt to establish a new order of mobile life where the cycle is no longer subordinate to the car" has been identified by many scholars as a missing element in planning for cycling in low-cycling and car-dominated national contexts, in which existing cultures, attitudes, behaviours, discourses and beliefs posit cycling as inherently alternative (Stehlin, 2014), marginal (Aldred, 2010), countercultural (Furness, 2005), foreign (Sirkis, 2000) or bearing stigma (Aldred, 2012). 'Soft' measures have already been found to be effective in reducing car travel (Bamberg et al., 2011). Bowles, Rissel and

Bauman (2006), in a study in Australia, found that mass community cycling events were key to attracting new riders.

As early as 2002, McClintock, in a chapter titled, 'Promoting cycling through "soft" (non-infrastructural) measures', proposed that infrastructure and cycle routes are neither a necessary nor a sufficient condition for high levels of cycle use: 'Attention must be given to other factors ... traffic speed, congestion ... and cultural attitudes...' (Jones [2001] in Mclintock, 2002b: 36) He describes 'softer' cycling policies as those that concentrate on the positive advantages of cycling, 'whereas infrastructure ... can be seen as having negative connotations, being implemented primarily to deal with the negative attributes of cycling ... with harmful consequences for increasing cycle use.'

In work conducted in the US, anthropologist Lugo (2013) expanded on the concept of soft infrastructure, adding 'human infrastructure' to the bicycling lexicon – meaning changed social attitudes in addition to changed built environments, brought about by projects such as group rides, social networks of activists, and the presence of bike commuters during rush hour (ibid.).

In guidelines and recommendations produced for or by other African cities, the role of 'soft' infrastructure has also been highlighted, albeit as a component of an infrastructure-first approach (see Rwebangira, 2001; Servaas, 2000). For example, a GTZ training course for NMT practitioners in African cities (Hook and Heyen-Perschon, 2003) devotes only a few of its 100 pages to the following two headings: 'Overcoming cultural barriers to [non-motorised vehicle] use' and 'Bicycling and walking promotion to increase use'. This is characteristic of similar texts, where reference of soft infrastructure tends to be considered only insofar as it might hinder the success of hard infrastructure. Furthermore, cultures, attitudes, behaviours, discourses and beliefs are seldom the target of long-term campaigns; often, it is assumed that the organization of promotional events will be sufficient to address all soft infrastructural challenges.

Bicycling policy and implementation nationally and in Cape Town

Policies and programmes to increase the rate of bicycle transport (or utility cycling) in South Africa place great emphasis on the need to create physical bicycling infrastructure. The first Engineering Guidelines for the Planning and Design of Pedestrian and Bicycle Facilities were published in South Africa in 1987. These guidelines were updated in 2003 (CSIR, 2003) and again in 2014, in order to 'provide guidance on the planning and design of safe pedestrian and bicycle facilities' (Vanderschuren et al., 2014). The new guidelines included examples of best-practice segregated interventions.

South Africa's first draft National Non-Motorised Transport (NMT) Policy (NDoT, 2008) states that 'the development of road networks and other transport infrastructure is a necessary condition to facilitate cycling', and proposes a Non-Motorised Transport Fund that would finance 'cycling-related road infrastructure improvement'. The updated national NMT Policy will no longer stand on its own, but is likely to become subsumed into the National Roads Infrastructure Policy, along with chapters on freight, roads funding and roads maintenance.[7]

The funding of infrastructure is further facilitated through the Public Transport Infrastructure and Systems Grant (PTISG), administered by the country's national Department of Transport, which invites local authorities to submit budget proposals for development and implementation of non-motorized transport facilities and infrastructure. Programme success is measured by the National Department of Transport (e.g., 2011/12–13/14[8]), as well as by the Provincial Government Western Cape (Transport for Cape Town, 2015) largely in terms of the number of kilometres of constructed bicycle lanes. In the Provincial Land Transport Framework (PLTF), a key objective is that dedicated cycle lanes in the Western Cape 'must be doubled by 2014' (Western Cape Government, 2012).

In Cape Town in the early 2000s, cycling was explicitly identified as a core component of a well-functioning transport system, with the publication of the Cape Town Metropolitan Bicycle Master Plan in 2002. Although the Plan was the first in South Africa to position cycling as a mode requiring specific attention from policy makers and planners, it implicitly positioned the promotion of cycling as synonymous with the construction and implementation of cycling infrastructure, and had an evident focus on recreational cycling.

Since 2005, with the publication of its Draft NMT Policy and Strategy (City of Cape Town, 2005) – which incorporated the Master Plan – the City of Cape Town has had a stated goal of using infrastructure to increase cycling in the city, 'by creating a safe and pleasant bicycle and pedestrian *network of paths*' (emphasis added). This Draft NMT Policy and Strategy represented the first major attempt to craft a coherent, comprehensive and specific vision of NMT's place within the city's transport mix, in which 'hard' and 'soft' infrastructure were differentiated. The document had four primary aims (emphasis added):

- Increase cycling and enable walking as modes of travel
- Create safe pedestrian and cycling environments
- Develop a quality, attractive and dignified environment
- Promote a changed *culture* that accepts the use of cycling and walking as acceptable means to move around in the city and elicit more responsible NMT behaviour

The Western Cape's Draft Strategy for NMT, published in 2009 (Provincial Government Western Cape, 2009), also identified infrastructure as key to attracting bicycle users, particularly those categorized as discretionary users:

These discretionary users require a high quality of services and *infrastructure*, and may revert to motorised travel if quality is not adequate.

NMT projects in the Provincial Strategy are described specifically in terms of hard infrastructure: sidewalks, cycle paths or lock-up facilities. Awareness, training and education are described as largely the domain of the private sector, non-governmental organizations or civil society. Although these early documents did recognize 'soft' as well as 'human' infrastructural requirements – including the need for bicycle promotion, bicycle counts, increased law enforcement, and the increased role of stakeholders in a broad range of NMT planning activities, and these concerns have been repeated in subsequent Comprehensive Integrated Transport Plans[9] (CITPs) for the City (Transport for Cape Town, 2015), the concrete outcomes of the policies have been infrastructural rather than behavioural.

During the 10 years following the adoption of the City's 2005 NMT Policy and Strategy, an ambitious programme of infrastructure provision was implemented, which would achieve a substantial degree of physical coverage of the city. This metro-scale programme to develop a comprehensive NMT network plan involved the appointment of teams of consultants for a three-year period for each of the four transport regions into which the city had been divided, each of which was tasked with prioritizing, designing and implementing NMT projects. At completion in 2012, 18 such projects had been implemented. At the time of writing (2016), the project was in Phase 2, with 12 projects under construction and seven out for public tender (pedestrian or shared sidewalk facilities).

However, only a fraction of this infrastructure provision involved high-quality, purpose-designed bicycle facilities or bicycle-only lanes. Instead, it focused on pedestrian infrastructure such as sidewalk construction, pedestrian refuges, pedestrian crossings, sidewalk upgrades, walkways or surfacing of sidewalks (Transport for Cape Town, 2015). In the 2013–2014 budget cycle, for example, of the 41 NMT projects in the plan (City of Cape Town, 2015), only one involved the construction of a bicycle lane. Phase 3 is set to commence in the near future. Chapter 8 raises a further concern, suggesting that Cape Town's bicycle (and pedestrian) infrastructure itself in many instances does not meet sufficient or best-practice standards (including standards proposed in the national facility guidelines), and that the bicycle infrastructure that does exist cannot be shown to have attracted cyclists.

At the same time, expenditure on NMT as a budget line item has continued to rise, with substantial expenditure allocated for financial 2015–2018 – in the order of R120 million every year (Transport for Cape Town, 2015). Total spending is about R1.4bn p.a. (City of Cape Town, 2015). Although 120 million might seem low in comparison to the total figrue, at least R1.2 bn is allocated to the MyCiTi Bus Rapid Transport system alone, thus a figure of 120 million puts NMT high on the spending list.

This increasing scale and scope of NMT infrastructural interventions was matched by the growing profile of NMT in administrative terms. By 2006, cycling had been recognized in the City's Integrated Transport Plan (ITP), and an NMT unit had been created within the Transport Directorate[10] (Kok, 2014). In 2010, a city-wide NMT Masterplan was completed, along with flag-ship NMT interventions related to the 2010 FIFA World Cup; in the follow-ing year, the Bicycle Masterplan was updated to better reflect actual cycling traffic based on input from cycling civil society (ibid.). By 2014, cycling had been recognized and supported within the city's Integrated Development Plan (IDP) and a new and more unified transport authority, Transport for Cape Town, was launched in 2012. The City's data capturing efforts also increased in complexity and ambition, with a total of 260 locations surveyed during the morning peak (06h00–08h30) since monitoring began in 2010 (ibid.). NMT is referenced several times in the most recent CITP, in a number of the chapters as well as the annextures; perhaps this further highlights the failure of policy to date, where NMT is mainstreamed but with little discernible change as an outcome.

Copenhagenizing Cape Town?

> The "build-it-and-they-will-come crowd" ought to realise there is no sil-ver bullet to increase cycling.
> (Marco te Brömmelstroet, Urban Cycling Institute, University of Amsterdam)

In their 2008 publication 'Making Cycling Irresistible: Lessons from the Netherlands, Denmark, and Germany', Pucher and Buehler suggest that infrastructure without extensive complementary behavioural and spatial interventions is unlikely to increase bicycle mode share. At the same time, Copenhagenization, or 'the Copenhagen model', has essentially become shorthand for international urban design and bicycle infrastructure best prac-tice. Numerous studies have focused on understanding the infrastructure and policy decisions made in Copenhagen, primarily over the past 35 years, in an attempt to condense the Copenhagen model into a global 'best practice' recipe of features that can be replicated anywhere in the world (Pucher and Buehler, 2008; Pucher et al., 2010; Servaas, 2000; Kåstrup, 2010).

This approach to some extent exemplifies the criticism of approaches to large infrastructure projects for other modes, such as Bus Rapid Transit implementation (Hitge, 2012; Salazar Ferro and Behrens, 2013, 2015; Lucas, 2011).

> Whereas BRT is nowadays a textbook example ... the development process of more than three decades is seldom acknowledged, and other cities tend to copy only the successful 'end-state'. (Hitge, 2012)

In Copenhagen, the capital city of Denmark, at least 45% of commuter trips are made by bicycle in greater Copenhagen, and 63% within Copenhagen (Copenhagen Together, 2014). The city has an extensive network of bicycle-only lanes (in other words, excluding pedestrians). Yet many of the characteristics that assure a high bicycle mode share are non-infrastructural. In contrast to sprawling Cape Town (Bruun et al., 2016), Copenhagen itself is dense and compact, with a flat topography; this may, in part, explain why 87% of bicycle trips are less than 5 km and only 4% are more than 11 km. Most trips are around 2 km on average (Denmark Technical University, n. d.).

Unlike in South Africa, helmets are not compulsory. Copenhagen also has one of the lowest car ownership rates in Europe with only 43% of families having a car (Statistics Denmark, June 26, Number 325, 2015). A new car is taxed at 180%, fuel is expensive and parking is difficult to find. Many streets are closed to car traffic, Dutch-style *woonerven* (car-free or 'home' zones where pedestrians and cyclists have priority) are abundant, and vehicle parking is often removed to create bicycle parking or lanes. Some of the major streets in the inner core are pedestrianized, and there are no highways through the central city. Busses and trains are abundant with frequent service, but they are typically slower than biking over short distances. As a small city of only 88 square km, cycling is almost always the fastest, cheapest and most convenient option for trips within the central city; owning a car isn't a status symbol and riding a bike isn't stigmatized.

Cape Town, on the other hand, is a city with enduring spatial, economic and demographic divides, all of which translate into its mobility practices (City of Cape Town 2014; Transport for Cape Town, 2015), with a vehicle-owning elite, and poor quality and infrequent public transport. It is not yet a city in which bicycling is able to compete – in terms of time, flexibility and convenience – with motorized transport (Bruun et al., 2016; City of Cape Town, 2015).

In Cape Town, bicycle advocacy is a relatively recent phenomenon (Jennings, 2010, 2015), whereas in Denmark bicycle activism and planning started more than a century ago. The Danish Cyclists Federation, the primary national advocacy group, dates to 1905, and cycling has since then always been part of transport planning. There has always been a voice for the cyclist – it has simply been louder or softer at different stages of history (Niels Jensen interview, 2010; Knudsen and Krag, 2005).

Copenhagen's 2015 vision strategy (published in 2009) includes nuanced baseline data, and goals that are specific and quantitative in nature to allow for easy evaluation; assessment of goal realization occurs on a biannual basis through the Bicycle Account. Cape Town's 2005 policy, by contrast, aimed simply to 'increase cycling'.

Bogotá, Colombia, the cycling exemplar from South America, is more comparable to Cape Town than Copenhagen is: Bogotá, too, is a place of significant economic and social inequity, increasing and rapid urban motorization, and long travel times, where paratransit vehicles play a crucial role in everyday transport.[11] As with Cape Town, in Bogotá NMT modes have the greatest share (walking), and there is a strong increase in the ownership and use of cars and motorcycles (Hidalgo and Huizenga, 2013). On the other hand, Bogotá is more like Copenhagen in that it is flat, dense and compact; three-quarters of daily trips in the city are less than 10 km and bicycles can often cover that distance faster than cars through the city's 'traffic-snarled streets' (Cervero, 2005).

In 10 years, from the mid-1990s to the mid 2000s, the share of daily cycling trips in Bogotá grew from 0.9 percent to 4 percent (Cervero, 2005). Like in Cape Town, NMT facilities are associated with the BRT service, but facilities extend well beyond the bus stations (Cervero, 2005). In 2009, Bogotá had more than 290 km of exclusive dedicated bicycle paths called *ciclorutas* (Massink et al., 2009; Montezuma, 2005). Cervero sums up Bogotá's bicycle transformation tactics thus:

> Bogotá is an extraordinary example of matching infrastructure 'hardware' with public-policy 'software': Latins America's most extensive network of cycleways, the world longest pedestrian corridor, and the planet's biggest Car Free Day (covering an entire city of 35,000 hectares). Today [2005], 43 per cent of the city's transport investment budget goes to ancillary policy measures.

Car and vehicle parking restrictions, through a license tag system and prohibition from Bogotá's central city streets during peak hours, are further key interventions (Cervero, 2005). This reinforces Buehler and Pucher's observation that the success of European cities has been to 'restrict car use and make it more expensive by establishing car-free zones and traffic-calmed neighbourhoods, while raising prices on parking' (Buehler and Dill, 2015, p.22). Pucher and Buehler (2012) offer a reminder that no single measure suffices: what is necessary is a network of integrated bikeways combined with car and parking restrictions, information, promotion, marketing to specific groups, education, motorist training, and enforcement. Bogotá thus constitutes in many senses a counterpoint to the narrative of motorization as an inevitable consequence of rising incomes in the developing world, although it is also unusual in that its concerted and high-level efforts to promote cycling date back to at least 1974 (Hoffmann and Lugo, 2014).

Build it and they don't come?

South African National Department of Transport officials Mashiri et al. (2013) have said that 'while funding for NMT infrastructure, principally for cycling and walking … is recognized, it does not include funding for promotional activities and the provision of [bicycles] – the assumption, which is not necessarily true, is that once the infrastructure is provided, it will be used.'

The City of Cape Town's approach bears this out, as there is as yet no clear evidence of a corresponding increase in overall rates of commuter cycling.

> 'Reports of rider attacked and mugged [again], cycle lane near Woodstock station'
> 'What's the point of a cycle lane if it can't be safely used?'
> 'I never use that portion of the cycle lane. I rather risk the [highway].'
> (Cyclist's comment on Twitter, July 2015, regarding the City of Cape Town's flagship MyCiTi cycleways, which the Integrated Transport Plan [2015 review] notes is the longest single bicycle lane on the African continent)[12]

In Cape Town, as with other cities and provinces in South Africa, it is the rollout of hard infrastructure that has been quantified, rather than a more holistic approach that takes both behaviour and facilities into account. The ongoing efforts of city planners to create a cyclable city routinely reflect a discourse in which the measured outputs are the numbers of bicycle lanes built (in kilometres) and the absolute number of cyclists crossing an intersection during morning peak hours. Changes in the modal share of cycling, the composition of the cycling public or other non-infrastructural factors have until recently remained largely unexamined.

This focus on infrastructure has been supported or actively called for by the bicycle advocacy community, much like that of the US, where 'bike advocates and researchers tend to emphasize urban form, often lobbying for northern European infrastructure models' (Pucher and Buehler, 2008; Lugo, 2013). For civil society, it may be that hard infrastructure is seen as visible political commitment. An incremental approach [gradually serving user needs] generally outlives political terms of office (Hitge and Van Dijk, 2012), 'but infrastructure more easily serves political ambitions, where implementation is easier to measure, and "launch", than is societal change'.

The City itself concedes that Cape Town's growing 'bicycle-friendly' culture is not of its making, but that of the 'community' (Transport for Cape Town, 2015, p. 57). This may be due to the fact that, until recently, the City offered minimal financial or bureaucratic support to events or programmes aimed at fostering a cycling culture (de Waal, 2015 personal interview; Tukushe, 2015 personal interview). Events that are widely held to have served as a catalyst for broad-based participation in cycling in several Latin American cities, such as the scheduled weekly exclusion of cars from major urban thoroughfares, have until relatively recently, been considered to be the preserve of volunteers rather than the City (Braake, 2013; de Waal, 2015 personal interview), and have been hampered by cost and bureaucratic burdens imposed by local government by-laws. Until 2011, bicycles were not permitted on any form of public transport at any time, and bicycle parking facilities remain unsafe, and few and far between.

This is not to suggest, of course, that bicycle lanes have no role to play in increasing bicycle mode share. But where hard infrastructure and built facilities are to play a role in encouraging more people to ride, these need to be of high quality, consistent, coherent and integrated with other modes (Pucher, 2008), rather than a few costly, showpiece 'legacy' projects (see also Chapter 8).

A recent analysis of the rapid development bicycle infrastructure in Seville, Spain (Marqués. et al., 2015) shows a significant modal increase as the result of 10 years of building segregated cycling infrastructure (the years 2006 to 2011). Key, however, was that the infrastructure formed a coherent, continuous network, paying particular attention to connectivity as well as what the authors describe as cohesion and homogeneity (i.e. the design of the bicycle lanes is similar throughout the network). The network in Seville followed highly visible main streets (to reduce the risk to personal safety), and avoided detours and multiple street crossings. More importantly, most bicycle facilities were built on previous parking or traffic lanes, thus helping to reduce vehicle traffic. The new infrastructure was soon complemented by a bicycle share service, which quickly added to the total bicycle daily trips.

Cape Town, in seeking to equitably distribute bicycle interventions throughout the four planning regions, has to some extent created ad hoc or disconnected pathways rather than route-based networks targeting the areas and routes in which commuter cycling can already be observed. User group engagement appears to have been limited to consultation with existing recreational cycling constituencies, rather than attempts to engage working-class cyclists commuting to industrial estates, which the city's own data suggest are the dominant cycling population in Cape Town (Transport for Cape Town, 2015, p. 31).[13] Existing cyclists are inadequately served by the unprecedented infrastructural investment of recent years.

Findings in Chapter 8 bear this out. Cycling infrastructure, which is sound in general, fails in the details, resulting in a fragmented network, multiple street crossings, and a multiplicity of design typologies (Beukes, 2011). There has been little implementation of ancillary infrastructure, such as a comprehensive city-wide signage and wayfinding system for cyclists, which has been shown to be an important component of successful NMT networks (AASHTO, 2012), although such a system is slated for implementation in the latest CITP. A small number of bicycle-signalling sets have been added at intersections (Kok, 2014).

Shifting gears: a new direction in bicycle planning?

Recognizing that attempts to 'increase cycling' to date have been inadequate, in early 2015 TCT called for proposals for a cycling-specific strategy, which was to develop an understanding of the barriers to increased bicycle use and identify evidence-based strategies to grow cycling in the metro area.[14]

Early presentations to stakeholders by the consulting team (11 February 2016) noted that TCT now recognized that a different approach was needed, and acknowledged that to date, it had focused 'on the provision of cycling infrastructure, with little improvement ... and a large investment for a poor result'. A core reason for poor levels of cycling, suggested the presentation, is the lack of an integrated approach to and designated responsibility for cycling, and it recognized that a significant threat to growing mode share is an inability to grasp the need for a structured, integrated and collaborative approach.

As Reid (2014), author of *Roads Were Not Built For Cars*, writes, unless car use is 'tackled', and the car still remains the easiest way to get from A to B, improved bicycle infrastructure is unlikely to attract large numbers of new bicycle commuters. Transport for Cape Town has begun to develop a key set of supporting policies that, by reducing the ease with which private vehicles are able to move, will cumulatively be better able to facilitate bicycle transport. For example, TCT's Travel Demand Management (TDM) Strategy (Transport for Cape Town, 2015) and Parking Policy (Transport for Cape Town, 2014) both recognize the need for vehicle and parking restrictions, warning of the introduction of appropriate disincentives to private car use (City of Cape Town, 2015). In her address at TCT's Congestion Summit, in November 2015, Cape Town Executive Mayor Patricia de Lille put it to residents that they would need to change their travel behaviour and attitude towards public non-motorized transport such as walking and cycling. Further, she noted that, 'without operational and behavioural change projects running alongside infrastructure intervention, we will not have a sustainable approach'.

Conclusion

Over the course of the last three decades, research on urban transport within the East African cities that form part of this book publication has highlighted the urgent need to provide for cycling as a mode of great potential for mobility (see for example Rwebangira, 2001; Pendakur, 2005 and more recently Nkurunziza et al. 2012a, 2012b, 2012c). Such modes, wrote Howe and Bryceson in 2000, 'have lost out in the competition for policy attention and funding'. And as chapters 1 and 2 in this publication have argued, these modes remain marginalized.

Cape Town has since at least 2005 dedicated substantial policy and funding attention to bicycle transport, and has become an emblematic success story to others on the continent, although cycling's modal share remains extremely low, at 0.5%. This chapter therefore cautions that cities wishing to learn from the Cape Town experience are advised to learn from the city's mistakes, too – as Cape Town itself is doing. The examples of Copenhagen and Bogotá suggest that social interventions, as well as supporting policies and interventions, are crucial to the success of physical interventions in the absence of an existing cycling culture. Neglect of these may thus be an important contributing factor to Cape Town's slow pace of modal change.

Despite a full decade since Cape Town's first public formulation of the new NMT paradigm, it is clear that considerations other than the existence of bicycle lanes will play a critical role in the transformation of the city into a 'safe and pleasant' bicycle environment. Efforts to promote cycling have attained insufficient returns to date, and run a high risk of continued failure without a commitment to a change in direction. The importance of soft-infrastructure interventions has been recognized before – in the many policy statements and provisions – yet seldom has soft infrastructure been integrated into the City's planning, budgeting, implementation or monitoring and evaluation programmes.

In Cape Town today, the overwhelming majority of people who own and ride bicycles do not use a bicycle as transport. Their stated reasons do include a lack of cycling facilities (Irlam, 2016), but in addition, they cite the cost of bicycles, long travel distances, personal safety concerns, theft, aggressive driver behaviour, topography, wind speed, lack of transport integration, and the need for a car during work hours (City of Cape Town, 2010). These barriers could be overcome not only with bicycle infrastructure investment but with more broad-based interventions, such as bicycle promotion, the provision of information and wayfinding, quality public transport, modal integration, car-share and shower facilities, driver education and bicycle-safety legislation.

Cape Town, faced with an absence of a mass cycling culture to date, has placed insufficient emphasis on social sciences approaches and fearless evaluation. The 2017 draft Cycle Strategy, as released for public comment in January 2017, offers little evidence of a change in direction – and its draft vision continues the narrative thread of claiming leadership in terms of bicycle planning.[15] But for those whose lives and livelihoods are at risk from current conditions on the roads, the term 'strategy' had implied a definitive move from the broad and aspirational to the measurable and specific. It was in this hope that utility cyclists had given the City a little leeway over its missteps and misadventures, since it was expected that this strategy would at last join up all the dots.

Rather than behaviour change programmes, legislative reform, and the reclamation of public space from cars in favour of bicycles, for example, the authority has opted to date to promote utility cycling overwhelmingly in physical, asphalt-and-concrete terms, regarding it as an essentially infrastructural and bicycle-ownership challenge. Without the evidence to direct the effective allocation of resources, and until the promotion of cycling moves beyond the fragmented provision of 'hard' infrastructure (and more recently, the promise of provision of bicycles), the constraints to growing bicycle mode share will remain intact, limiting both the amount of cycling and the diversity of the cycling population.

Acknowledgements

With much gratitude to Dr Rahul Jobanputra at Transport for Cape Town, and Gerhard Hitge, for insightful comments on numerous drafts of this chapter.

Notes

1 Speech by Executive Mayor, Alderman Dan Plato at launch of the MyCiTi BRT-associated NMT lanes, media release 'Cape Town is Cycle City', February 2011. In Cape Town's Integrated Transport Plan ([ITP], City of Cape Town, 2015,

p. 137, item 8), an extract reads: 'Cape Town is renowned as the most bike-friendly city in Africa. It received an award in 2012 for the longest continuous cycle lane in Africa. The lane is 16.4 kilometres long.'

2 For example, Cape Town is claimed to be one of the world's top cities when it comes to urban cycling paths. www.southafrica.net/za/en/articles/entry/article-cape-town-urban-cycling-paths (accessed 15 February 2016)

3 Known as the MyCiTi service.

4 In South Africa, walking, cycling, skateboarding and animal-drawn transport are referred to as non-motorized transport (NMT). This chapter is concerned with bicycle transport (utility cycling) only, although the majority of policies, plans and strategies refer to NMT as a collective entity, and much of the data is collected and reported as a single mode.

5 Transport for Cape Town (TCT) was founded in 2012. Thus much of the NMT policy and implementation direction was taken before the establishment of the Transport Authority. This chapter therefore at times refers to the City of Cape Town as the relevant authority.

6 Name is known to the author.

7 To the best of the authors' knowledge at the time of writing, June 2016.

8 Baseline: three cities and three district municipalities (DMs) developed non-motorized transport (NMT) master plans and completed phase one of the pedestrian and bicycle tracks (constructed 10 km of bicycle lanes); KPI NMT infrastructure and facilities rolled out to municipalities; drive skills development programmes and reprioritizing the budget for NMT infrastructure development.

9 As prescribed by the national Department of Transport.

10 Author (Brett Petzer) communication with Teuns Kok, TCT, 19 June 2015, on the subject of New Commuter Cycling Infrastructure in the City of Cape Town, 2005–2015.

11 The average travel time in Cape Town was, at about 90 minutes in 2013, at the upper end of the global range, which averages around 70 minutes per person per day (Metz, 2010; Schafer and Victor, 1998). Of greater significance is the discrepancy between modes in Cape Town, with car users travelling at the global average of 70 minutes, but public transport users averaging around 110 minutes (City of Cape Town, 2013). This significant difference resonates with the discrepancy in the levels of spending on infrastructure for the two largely separate sub-systems of private and public transport networks, over the past three decades.

12 On page 137, item 8, the extract reads: 'Cape Town is renowned as the most bike-friendly city in Africa. It received an award in 2012 for the longest continuous cycle lane in Africa. The lane is 16.4 kilometres long. In order to build on the City's already established cycling culture, TCT will shortly be launching Cape Town's first bike share project.'

13 On page 31 of the 2015 Review of the CITP, the highest cycling counts among whole-day passenger trips, relative to the size of the area, are recorded inbound and outbound from industrial areas, such as Epping or Wynberg.

14 A tender for an update of the 2005 NMT Policy and Strategy was also awarded in 2015.

15 The vision of the Draft Cycling Strategy for the City of Cape Town is 'To make Cape Town the premier cycling city in South Africa' (draft January 2017).

References

AASHTO. 2012. Guide for the development of bicycle facilities, 4th Edition. Washington, D.C.:American Association of State Highway and Transportation Officials.

Aldred, R. 2010. "On the outside": constructing cycling citizenship. *Social and Cultural Geography*, 11(1):35–52.

Aldred, R. 2012. Incompetent or too competent? Negotiating everyday cycling identities in a motor dominated society. Mobilities. 8(July 2015): 37–41.

Bamberg, S., Fujii, S., Friman, M., and Garling, T. 2011. Behaviour theory and soft transport policy measures. *Transport Policy, 18*(1), 228–235. https://doi.org/10.1016/j.tranpol.2010.08.006

Bechstein, E. 2010. Cycling as a supplementary mode to public transport : A case study of low income commuters in South Africa. 29th Southern African Transport Conference (SATC 2010), 33–41. Accessed at: http://repository.up.ac.za/bitstream/handle/2263/14739/Bechstien_Cycling(2010).pdf?sequence=1.

Beukes, E. A. 2011. *Context-sensitive road planning for developing countries*. University of Cape Town.

Bowles, H., Chris, R., and Bauman, A. 2006. Mass community cycling events: who participates and is their behaviour influenced by participation. *International Journal of Behavioral Nutrition and Physical Activity*.

Braake, S. J. ter. 2013. The Contribution of Cicloruta on Accessibility in Bogotá: A spatial analysis on Cicloruta's contribution to job accessibility for different social-economic strata, Bachelor thesis, Civil Engineering, University of Twente/Los Andes University.

Bruun, E., Del Mistro, R., Venter, Y., and Mfinanga, D. 2016. The state of public transport systems in three Sub-Saharan African cities. In Behrens, R. Behrens, R., McCormick, D., and Mfinanga, D. (eds), *Paratransit in African Cities*, chapter 2. London: Earthscan.

Buehler, R., and Dill, J. 2015. Bikeway Networks: A Review of Effects on Cycling. *Transport Reviews, 1647*(May), 1–19. https://doi.org/10.1080/01441647.2015.1069908

Cervero, R. 2005. Accessible Cities and Regions: A Framework for Sustainable Transport and Urbanism in the 21st Century

City of Cape Town. 2005. *NMT policy and strategy volume 1: status quo assessment*. Cape Town.

City of Cape Town. 2005. *NMT policy and strategy volume 2: policy framework*.

City of Cape Town. 2010. City-wide NMT: South - strategic framework.

City of Cape Town. 2013. *Household survey report – final draft*. Cape Town: City of Cape Town.

City of Cape Town. 2014. *State of Cape Town 2014: celebrating 20 years of democracy*, 240. Pretoria: City of Cape Town.

Copenhagen Together. 2014. *Copenhagen City of Cyclists: The bicycle account 2014*. Copenhagen, Denmark: City of Copenhagen, The Technical and Environmental Administration, Mobility and Urban Space. Available at: www.cycling-embassy.dk/wp-content/uploads/2015/05/Copenhagens-Biycle-Account-2014.pdf.

Cox, P. 2013. Mass culture, subcultures and multiculturalism: how theory can help us understand cycling practice. In unpublished conference paper given at Velo-City 2013. Vienna, 11–14. Accessed at: http://hdl.handle.net/10034/317348.

CSIR. 2003. Pedestrian and Bicycle Facility Guidelines: Engineering manual to plan and design safe pedestrian and bicycle facilities (Draft 1.0). Pretoria: Council for Scientific and Industrial Research.

Denmark Technical University. n. d. Danish National Travel Survey. Lyngby, Denmark: DTU Management Engineering, Data and Model Centre, Transport DTU. Available at: http://www.modelcenter.transport.dtu.dk/english/tu/hovedresultater.

Furness, Z.M. 2005. 'Put the fun between your legs!': The politics and counterculture of the bicycle. PhD Dissertation, University of Pittsburgh, Pittsburgh, PA.

Goodman, A., Sahlqvist, S., and Ogilvie, D. 2013. Who uses new walking and cycling infrastructure and how? Longitudinal results from the UK iConnect study. *Preventive Medicine*, 57: 518–524.

Gwala, S. 2007. Urban non-motorised transport (NMT): a critical look at the development of urban NMT policy and planning mechanisms in South Africa from 1996–2006. Proceedings of the 26th Southern African Transport Conference. Pretoria, South Africa

Handy, S., van Wee, B., and Kroesen, M. 2014. Promoting cycling for transport: research needs and challenges. *Transport Reviews*, 34(1): 4–24.

Hidalgo, D., and Huizenga, C. 2013. Implementation of sustainable urban transport in Latin America. *Research in Transportation Economics, 40*, (1): 66–77 Available at: www.sciencedirect.com/science/article/pii/ S0739885912001060

Hitge, G., and Vanderschuren, M. 2015. Comparison of travel time between private car and public transport in Cape Town. *Journal of the South African Institute of Civil Engineering*, 57(3), 35–43.

Hoffmann, M.L., and Lugo, A.E. 2014. Who is "world class"? Transportation justice and bicycle policy. *Urbanities*, 4(1):45–61.

Hook, W., and Heyen-Perschon, J. 2003. Non-motorised transport in African cities – options for interventions and networking in African cities. Preserving and Expanding the role of NMT, Sustainable transport: a sourcebook for policy makers in developing cities, Module 3d, 2003, Walter Hook, GTZ, Eschborn, Germany.

Howe, J., and Bryceson, D. 2000. Poverty and urban transport in east transport in East Africa. World Bank.

Irlam, J. 2016. Barriers to cycling mobility in Masiphumelele, Cape Town: a best-worst scaling approach. A dissertation submitted in fulfillment of the requirements for the degree of Master of Philosophy (Climate Change and Sustainable Development). University of Cape Town, South Africa.

Jennings, G. 2010. The visibility and purpose of cycling in Cape Town, South Africa. Paper presented at Bicycle Politics Workshop, Centre for Mobilities Research, University of Lancaster, September 2010.

Jennings, G. 2014. Finding our balance: considering the opportunities for public bicycle systems in Cape Town, South Africa. *Research in Transportation Business and Management*, 15: 6–14.

Jennings, G. 2015. A bicycling renaissance in South Africa ? Policies, programmes and trends in Cape. *Proceedings of 34th Southern African Transport Conference* (SATC 2015), 486–498.

Jobanputra, R. 2013. Vehicle-pedestrian and infrastructure interaction in developing countries. Thesis, Doctor of Philosophy in Civil Engineering, University of Cape Town.

Kåstrup, M. 2011. The cycling girl in Copenhagen and beyond, Velocity Sevilla. Presentation by Marie Kastrup at Velo-city, Seville, 2011.

Kåstrup, M. 2010. Til en kulturel cykelfaglighed, Trafik og vej. Accessed at: http://asp. vejtid.dk/Artikler/2010/01/7775.pdf.

Knudsen, W., and Krag, T. 2005. På cykel i 100 år: Dansk Cyklist Forbund 1905–2005. Danish Cyclist Federation

Kok, T. 2014. Cycling in the City of Cape Town – status quo and developments. Powerpoint Presentation given 12 June 2014.

Krizek, K.J., Barnes, G., and Thompson, K. 2009. Analyzing the effect of bicycle facilities on commute mode share over time.*Journal of Urban Planning and Development*.

Labuschagne, K., and Ribbens, H. 2014. Walk the talk on the mainstreaming of NMT in South Africa. 33rd Southern African Transport Conference, Pretoria.

Lucas, K. 2011. Applying a social exclusion approach to transport disadvantage in South Africa. THREDBO 12 Workshop 6.

Lugo, A. 2013. CicLAvia and human infrastructure in Los Angeles: ethnographic experiments in equitable bike planning. *Journal of Transport Geography*, 30: 202–207.

Marqués, R., Hernández-Herrador, V., Calvo-Salazar, M., and García-Cebrián, J.A. 2015. How infrastructure can promote cycling in cities: Lessons from Seville. *Research in Transportation Economics*, 53: 31–44.

Massink, R., Zuidgeest, M., Rijnsburger, J., Sarmiento, O. L., and Van Maarseveen, M. 2011. The climate value of cycling. *Natural Resources Forum*, 35(2), 100–111.

Mashiri, M., Maphakela, W., Chakwizira, J., and Mpondo, B. 2013. Building a sustainable platform for low-cost mobility in South Africa. 32nd Southern African Transport Conference, Pretoria.

McClintock, H. 2002a. *Planning for cycling: principles, practice, and solutions for urban planners*. CRC Press. Retrieved from: http://books.google.com/books?id=17-drWGaTQkCandpgis=1.

McClintock, H. 2002b. Promoting cycling through 'soft' (non-infrastructural) measures. In McClintock, H. (ed), *Planning for cycling: principles, practice and solutions for urban planners*. CRC Press (Taylor and Francis Group).

Montezuma, R. 2005. The transformation of Bogota, Colombia, 1995–2000: Investing in citizenship and urban mobility. *Global Urban Dev. 1* (1), 1–10.

National Department of Transport (NDoT). 2008. *(Draft) Non-Motorised Transport Policy.*

Nkurunziza, A., Zuidgeest, M., Brussel, M., and Maarseveen, M. 2012a. Examining the potential for modal change: motivators and barriers for bicycle commuting in Dar-es-Salaam. *Transport Policy*, 24(2012): 249–259.

Nkurunziza, A., Zuidgeest, M., Brussel, M., and Van Maarseveen, M. 2012b. Key events and their effects on cycling behaviour in Dar-es-Salaam. In 5th International Conference on Traffic and Transport Psychology, ICTTP 2012, 29-31 August 2012, Groningen, the Netherlands (pp. 75–75).

Nkurunziza, A., Mark Zuidgeest, M., and Van Maarseveen, M. 2012c. Identifying potential cycling market segments in Dar-es-Salaam. *Tanzania Habitat International*, 36(2012): 78–84.

Pendakur, V.S. 2005. Non-motorised transport in African cities: lessons from experience in Kenya and Tanzania, SSATP Working Paper No.8.

Provincial Government Western Cape. 2009. *NMT in the Western Cape draft strategy.*

Provincial Land Transport Framework. 2012. Western Cape Government. Accessed at: www.westerncape.gov.za/general-publication/provincial-land-transport-framework-2012.

Pucher, J., and Buehler, R. 2008. Making cycling irresistible: Lessons from the Netherlands, Denmark and Germany. *Transport Reviews, 28*(4), 495–528. https://doi.org/10.1080/01441640701806612

Pucher, J., Dill, J., and Handy, S. 2010. Infrastructure, programs, and policies to increase bicycling: An international review. *Preventive Medicine*, 50, S106–S125.

Pucher, J. and Buehler, R. 2012. *City cycling*. Boston: MIT Press.

Ribbens, H., and Gamoo, L. 2006. The need to provide safe and secure non-motorised transportation infrastructure and amenities. African Transport Conference. Pretoria, South Africa.

Reid, C. 2014. *Roads were not built for cars*. Island Press.

Rwebangira, T. 2001. Cycling in African cities: status and prospects. *World Transport Policy and Practice*, 7(2): 7–10.

Salazar Ferro, P., and Behrens, R. 2013. Paratransit and formal public transport operational complementarity: imperatives, alternatives and dilemmas. 13th World Conference on Transport Research, 15–18 July 2013, Rio de Janeiro, Brazil.

Salazar Ferro, P., and Behrens, R. 2015. From direct to trunk-and-feeder public transport services in the urban south: Territorial implications. *Journal of Transport and Land Use*, 1: 123–136.

Schafer, A. and Victor, D. 2000. The future mobility of the world population. *Transportation Research Part A 34*: 171–205.

Servaas, M. 2000. *The significance of NMT for developing countries*. I-ce, Interface for Cycling Expertise, Utrecht, the Netherlands.

Sirkis, A. 2000. Bike networking in Rio: the challenges for non-motorised transport in an automobile-dominated government culture. *Local Environment*, 5(1):83–95.

Stehlin, J. 2014. Regulating inclusion: spatial form, social process, and the normalization of cycling practice in the USA. *Mobilities*, 9(1):21–41.

Transport for Cape Town. 2013. *Road safety strategy for the City of Cape Town 2013-2018*. Cape Town.

Transport for Cape Town. 2014. *Integrated public transport network 2032: network plan*. Cape Town.

Transport for Cape Town. 2015. *Comprehensive integrated transport plan 2013–2018: 2015 review*. Cape Town.

Transport for Cape Town. 2015. *Comprehensive Integrated Transport Plan 2013–2018*. Pretoria: City of Cape Town.

Vanderschuren, M., Phayane, S., Taute, A., Ribbens, D.H., Dingle, N., Pillay, K., Zuidgeest, M., Enicker, S., et al. 2014. National Department of Transport NMT Facility Guidelines. (September 2015). National Department of Transport, Pretoria, South Africa.

Van Dijk, G and Hitge, E. 2012. Incremental approach to public transport system improvements. SATC 2012. Accessed at http://citeseerx.ist.psu.edu/viewdoc/download?doi=10.1.1.921.7494&rep=rep1&type=pdf

Western Cape Department of Transport and Public Works. 2010. *Draft non-motorised transport in Western Cape strategy*. Cape Town.

Western Cape Government. 2012. *Provincial Government Western Cape Land Transport Framework 2012*.

14 Grounding urban walking and cycling research in a political economy framework

Meleckidzedeck Khayesi, Todd Litman, Eduardo Vasconcellos and Winnie V. Mitullah

Introduction

Experience and research show that politics and economics significantly influence the development and implementation of walking and cycling policy, as well as transport policy, in general (Flyvbjerg, 2002, 2007, 2009; Garrison and Levinson, 2006; Kane, 2011; Pellegrino, 2011; Klopp, 2011; Koglin and Rye, 2014). However, political and economic factors are not, generally, acknowledged and included in the predominantly positivist transport research approaches, such as the four-stage transport-land-use model, land-use and transport integration model, multi-modal transport planning approach, and statistical modelling of travel behaviour (Vasconcellos 2001; Knoflacher 2009; Goetz et al., 2009; Dimitriou, 2011; Koglin and Rye, 2014). The consequence of the above omission is that the political economy issues that affect transport policy are less analyzed within the conventional transport research approaches.

It is a contradiction in terms that transport infrastructure development and policies that are influenced by political decision making continue to be mainly analyzed from a positivist approach. Statistical modelling is a well-established approach in transport research and there are even courses on it, as well as transport researchers who focus on this approach. This focus contrasts sharply with limited focus on a political economy approach that is rarely offered as a specialized framework or technique in transport courses. There is substantial knowledge on technical transport analysis and solutions but limited information on the political and economic context that may influence the uptake of these technical solutions. This gap in knowledge is well summarized by Vasconcellos (2001):

> A large body of knowledge has been developed, used by traffic engineers all over the world, utilizing quantitative techniques based on street capacity, vehicle dimensions and human physical characteristics, to decide how street space will be distributed among the users (Pignataro, 1973; Institute of Transportation Engineers, 1965; Transportation Research Board, 1985; Texas Institute of Technology, 1996). Implicit is the idea that this technical division is neutral and allocates equal benefits to

everybody. Also implicit is the assumption that political analysis of the conflicts, related to the use of streets, does not pertain to traffic management and should be treated elsewhere. It follows that the analysis of the use of streets has to be the exclusive domain of traffic engineers and their technical procedures. This way of thinking stresses quantitative approaches and underestimates, or disregards completely, the social and political aspects of urban circulation. It becomes especially problematic in developing countries where the political and social conditions differ widely from those in the developed world and where traffic environments are much more complex (Pignataro, 1973; Institute of Transportation Engineers, 1965; Transportation Research Board, 1985; Texas Institute of Technology, 1996).

What is this chapter about? The chapter explains a political economy theoretical framework that can be used for analysing walking and cycling travel behaviour and policy response in urban transport research in Africa. The chapter does not develop a new framework but rather draws on existing political economy knowledge to provide a conceptual model for analysing the political economy of walking and cycling in urban transport research. The chapter uses this framework to highlight its implications for a political economy analysis of walking and cycling in urban Africa.

Methods

This chapter is based on a review of literature and elaboration of a political economy framework for walking and cycling in urban research. This literature review focused on trends in transport research approaches. It identified a gap in the theoretical basis of this research, namely, inadequate attention being given to the political economy of walking and cycling.

The next stage was to derive and elaborate a political economy conceptual model. The derivation of the model consisted of examining the literature on political economy theory and identifying key concepts and constructs. The concepts and constructs were then used to elaborate a political economy model relevant for urban walking and cycling research. The concepts and constructs are described and possible research questions indicated. The model, presented later in this chapter, offers a theoretical basis for researchers to formulate propositions and hypotheses for current and future research. The chapter also shows how this book has addressed some of the research questions raised.

What is political economy?

Political economy has a fairly long history, which has led to the development of four main schools of thought, or perspectives, that currently contend for prominence: classical liberalism, radicalism, conservatism and modern liberalism (Clark, 1998; Liodakis, 2010). These schools reflect the importance of underlying ideologies and values in the analysis of political economy. The different schools have different orientations on how politics and economics should be approached in society, the place of the individual, community, equality and hierarchy. It should be noted that there are also variations in each broad group.

What is important about the political economy approach is that it enables an analysis of the primary goals being pursued in society, the institutional arena within which the goals are pursued and the primary actors who choose the goals from both economics and politics (Clark, 1998). The market and government are treated as the institutional arena. The government and market are seen as both political and economic institutions. Further, political economy considers the class structure of each society, power and decision-making, production of goods and services, and distribution of resources within a historical and contemporary context (Liodakis, 2010). It can be seen that a political economy approach goes beyond the easily observed features of society to investigate the underlying political and economic structures and relationships. It employs both qualitative and quantitative techniques, drawing on theories and concepts from diverse disciplines such as Economics, Political Science, Sociology, Geography, Environmental Science, Public Health, History and International Relations.

As pointed out by the Department for International Development (2009, p. 1), by bridging the traditional concerns of politics and economics, political economy 'focuses on how power and resources are distributed and contested in different contexts, and the implications for development outcomes. It gets beneath the formal structures to reveal the underlying interests, incentives and institutions that enable or frustrate change. Such insights are important to advance challenging agendas around governance, economic growth and service delivery, which experience has shown do not lend themselves to technical solutions alone.'

While the different schools of thought and individual researchers may pay attention to specific issues and geopolitical settings, the central concern of a political economy approach is the examination of the interaction of political and economic processes in a society, focusing on the distribution of power and wealth between different groups and individuals, and the processes that create, sustain and transform these relationships over time (Department for International Development, 2009). A political economy approach serves both academic and planning purposes. Whereas it improves our understanding of the political and economic processes occurring in society, this knowledge may be used in decision-making, for example, on actions that may be taken to address transport inequality and social exclusion, if the analysis shows extreme variation in the provision of transport infrastructure and services (Lucas and Currie, 2011; Ahmed et al., 2008). The political economy approach can be applied to different issues at varying geographical and administrative units, such as health reform at the national level (Fox and Reich, 2013; Reich, 2002), power dynamics in compliance with the Convention on Biodiversity at the global level (Boisvert and Vivien, 2012), politics of sustainable development at the global level (Liodakis, 2010), and structural causes of under-development at a continental and national level (Galeano, 2009; Leys, 1975). It is within this context that the next section shows how this approach may be applied to analysing walking and cycling policy in urban Africa.

The conventional transport research approach

The conventional transportation research and planning process begins with travel surveys that collect information on travel activity, which is used to estimate travel demand (Mees, 2009; Dimitriou, 2011). Such surveys tend to undercount walking and cycling travel, because they often under-count shorter trips, off-peak trips, non-work trips, travel by children and recreational travel (Stopher and Greaves, 2007, Alliance for Biking and Walking, 2010). They generally ignore the walking and cycling part of motor vehicle trips. For example, a bike-transit-walk trip is usually coded simply as a transit trip, and a motorist who walks several blocks from their parked car to a destination is simply considered an automobile user. Future traffic conditions are predicted using a four-step model, based on the following sequence: trip generation, trip distribution between origins and destinations, mode share (which generally does not include non-motorized transport trips) and route assignment.

These modelling results are used to predict future traffic trends, which are then used to prioritize transport system improvements. Transport planning agencies often produce maps showing current and projected future peak-period vehicle traffic speeds and roadway level of service ratings. This process seems logical and rational, but it is actually incomplete and biased for the following reasons (Hook, 1994; Vasconcellos 2001; Mees, 2009; Litman, 2013):

- Transport statistics significantly under-report walking and cycling travel activity, which leads to an incomplete picture of travel behaviour.
- It provides little information on demand for alternative modes, such as how walking and cycling improvements, or land-use policy changes, would affect travel activity. It often gives little consideration to the travel demands of physically, economically and socially disadvantaged people who rely primarily on walking and cycling.
- It reflects motorized travel or mobility, rather than accessibility-based planning and, therefore, fails to consider other factors that affect accessibility, such as the quality of alternative modes and land-use development patterns, as well as the politics of transport policy. It fails to reflect trade-offs between different types of accessibility, such as when wider roads and increased traffic speeds reduce pedestrian access and, therefore, public transit access, since most transit trips involve walking links. Better service for motorized vehicles often means increased barriers and road safety issues for walking and cycling.
- It only quantifies a limited set of impacts, which typically include travel time, vehicle operating costs, crash and emission rates. It tends to overlook

other important impacts, including parking costs, vehicle ownership costs, noise costs, barrier effect costs and sprawl-related costs.

• It fails to account for social equity objectives, such as the quality of accessibility options for physically and economically disadvantaged people, and affordability. By favouring automobile travel over more affordable modes, conventional planning imposes excessive costs on low-income people. The evaluation process is mechanistic and difficult to understand. People who use the analysis results often have little idea of their omissions, biases and uncertainties, nor the tendency of such planning to encourage automobile travel, to the detriment of other modes. Affected people often have little opportunity to influence decisions. Information about the planning process and opportunities for stakeholder involvement are often limited, particularly for physically, economically and socially disadvantaged people.

Many transport professionals now realize the limitations of the conventional transport approaches (Docherty and Shaw, 2011), and are partly contributing to prioritizing walking and cycling in an efficient and equitable transport system by conducting research (Legacy et al., 2012; Khayesi et al., 2010) and supporting or implementing specific policies and projects, such as complete streets, multi-modal planning, traffic speed control and walkable neighbourhood planning (Whitelegg, 2011; Mees, 2009, 2010). New evaluation tools, supporting this type of planning, have also been developed and are being used (Litman 2013). However, it is important to note that developing and using these new transport planning tools does not necessarily address the political economy issues related to walking and cycling. As Wachs (1995, p. 270) has pointed out: 'Transportation decisions are made in the political arena, and no analysis of travel and transportation networks is complete without the political dimension.' This observation echoes an earlier observation made about 50 years ago by Wolfe (1963) on the intricate relationship between transport and politics, including the development of transportation technology, military/strategic decisions and political management and consolidation of nation states. Strangely, this small and informative book is rarely cited in contemporary transport publications or included in reading lists for transport courses.

A political economy model for urban walking and cycling research

A framework for analysing the political economy of walking and cycling is presented in Figure 14.1. The model shows that quality of walking and cycling is an outcome of the interaction between urban transport and urban governance systems. An analysis of features of walking and cycling, such as trip purpose, trip length, mode use, trip origin and destination, safety, accessibility

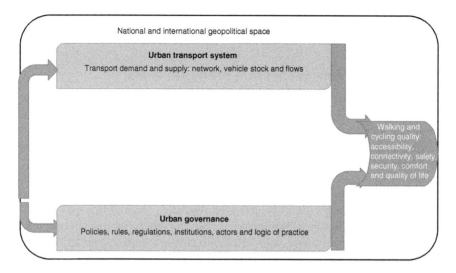

Figure 14.1 Situating walking and cycling research in a political economy model.

and provision of infrastructure for these two modes is important and a good starting point to lead to further analysis of decision-making and the planning of these modes of transport, within the context of urban transport systems, urban governance and the national and international geopolitical development agenda.

A key feature in Figure 14.1 is that walking and cycling are situated within, and influenced by, multiple layers of political arenas and decision-making processes operating at local urban, national, continental and international levels. It shows that planning for walking and cycling in urban areas involves decisions and actions by different actors and institutions within and outside the transport sector. It is within these layers that decisions are made on investing in walking and cycling modes of transport, transport in general, and other sectors of the economy. What this framework may imply to researchers is that they need to move beyond concentrating on the analysis of travel behaviour, network structure and flows of the transport system to examining decision-making processes and the extent to which options in favour of walking and cycling are implemented, or why there is a lack of implementing them despite political statements in favour of these modes. These topics are important and need to be analyzed but the way they are approached, using the positivist approach, leaves out relevant information on dynamics of transport policy decision-making and overall governance of the transport sector and society. Thus, applying a political economy framework may contribute to providing insights into the dynamics of decision-making and identify some leverage points for intervention in both the urban political context and transport planning process.

Using Figure 14.1, a number of political economy questions, regarding walking and cycling in urban Africa, can be raised and explored with respect to the following themes:

What is the travel behaviour related to walking and cycling?

Understanding levels and conditions under which walking and cycling take place is important for planning. Some studies have been conducted on this aspect in African cities (Behrens, 2005; Venter et al., 2007, Salon and Gulyani, 2010; Salon and Aligula, 2012). This book has consolidated this research and provided in-depth analysis of travel behaviour related to walking and cycling, including the risk of road traffic injuries, and assessment of infrastructure for walking and cycling in Nairobi and Cape Town (see Chapters 2–6). However, there are several issues yet to be analyzed regarding the multiple uses (accessibility, mobility, business, residence, recreation, sports, meetings, conservation and political demonstrations) of urban transport space, regulations and strategies utilized to guarantee the multiple uses, competition and struggle for this space, and quality of this space. Other topics that have been inadequately researched are access and mobility related to urban production and distribution systems, rural–urban interaction, adoption and use of cell phone and information and communication, in general, as complementary and alternatives to physical movement, equity in access and mobility among different social groups, walking and cycling within physical activity behaviour of urban residents, and work and non-work travel behaviour. In short, a comprehensive analysis of travel behaviour, mobility and access in urban Africa is needed and it is important that researchers and donors invest time and resources in this type of research. Unlike rural transport in Africa, which is now being synthesized in form of a literature review (Porter, 2014), urban travel behaviour research remains scattered and requires a comprehensive synthesis of the existing evidence to inform current and future research.

Who cares about walking and cycling in policy-making processes?

For walking and cycling to receive the support they need, political leaders, policy makers, practitioners, researchers and advocates who are moving action on these modes will need to build understanding and political support. In many situations, low levels of walking and cycling result from past and current policies that favour motorized over walking and cycling travel through strategies such as under-investment in sidewalks and bike paths, promoting roadways designed to maximize traffic speeds, and implementing zoning codes that stimulate sprawl (Monheim, 1996; Litman, 2013; Koglin and Rye, 2014). In many situations, policies that favour motorized travel result from the greater political influence of residents and businesses that use motorized modes, compared with people who rely on walking, cycling and public transport. Perhaps a starting point may be an examination of decisions made within and outside the transport sector, tracing urban and national development issues and showing how these have affected decision-making in transport and the structure and development of existing urban transport systems. Such an analysis has the capacity of not only highlighting internal factors but also external influences from donors, consultants, foreign governments and global discourses that affect decision-making in developing countries. The analysis may use detailed case studies of sectors, such as public passenger services and development of urban transport infrastructure (road, railway and airports) to show how various decisions made have affected investment in, and development of, these modes that have been neglected and yet are important to the poor majority. This kind of analysis would provide insight into politics and governance, political processes, location of power and decision-making in urban transport in developing countries.

This book has raised issues and also offered some answers to questions related to land-use planning, urban development and institutions in Chapters 7, 8, 12 and 13. However, central research questions requiring further analysis may be: (a) How have decisions and practices made at different times, influenced by political, social and economic considerations from within and outside developing countries, affected the planning and development of urban transport, in particular, walking and cycling in urban Africa? (b) How have these decisions and practices specifically affected accessibility, infrastructure and services (e.g. resting places, parking lots) for walking and cycling? (c) How do transport planning institutions, professionals and other actors perceive and approach planning for walking and cycling versus motorized modes of transport?

Workings of local and national political systems, and the influence of external and internal factors on the capacity of the state

Planning tools and guidelines on how to incorporate walking and cycling into urban transport planning in Africa, and other parts of the world, exist (Stucki, 2015). There are also technical solutions that, if implemented, would ensure that walking and cycling are catered for in policy (Stucki, 2015). Further, there is evidence showing the environmental, health and economic benefits of walking and cycling, as well as issues that require a policy response (Black et al., 2016). However, the practical reality is that walking and cycling remain neglected in transport policy around the world. This state of affairs may lead us to ask: Why is motorized transport favoured over non-motorized modes, even in developing countries, when evidence exists on the benefits of these modes of transport, and tools are available to help plan for them?

Transport systems, including walking and cycling, are part of the services provided by the state and other agencies. Prioritizing walking and cycling by the state is a contested policy process. The policies pursued by the government in power, at both national and local level, have an effect on transport policy and practice (Tyler, 2004; Kane, 2011; Docherty and Shaw, 2011).

If the focus of the state policy is on motorized transport, then walking and cycling will receive limited priority. This book has shown some aspects of the role of the state in transport policy formulation and implementation in Chapters 12 and 13. However, there is a need for further research to answer the question: What does it take politically, institutionally and economically to prioritize walking and cycling in urban development and transport policy? A study on urban passenger transport in South African cities has highlighted the dynamic institutional and planning issues that are relevant to answering the above question, which researchers working in other African cities can draw upon (Wilkinson, 2008). The ongoing road and transport infrastructure development in African countries and cities provides a vital setting to conduct research on decision-making on inclusive facilities for modes, including walking and cycling. For example, understanding the reasons for omission of facilities for walking and cycling can be examined by conducting case studies of recent highway improvements that have not catered for these modes. This omission could be due to cost overruns and political issues like rent seeking that are associated with infrastructure projects (Flyvbjerg, 2005).

Conclusion

The central argument of this chapter is that urban transport research should not focus mainly on statistical assessment of walking and cycling, as it is commonly done in network and travel behaviour analysis, but also seek to examine walking and cycling from a political economy perspective. This argument is motivated by the real-world experience showing that the agency of the political and economic environment at local, national and international levels plays a key role in translating transport knowledge into practical solutions in the form of complete streets, active design guidelines, safe walking and cycling environments, an integrated public transport system, and allocation of human and financial resources to planning for cycling and walking. The chapter has drawn on the political economy theory to provide a framework for analysing the political economy of walking and cycling in urban transport research.

This chapter has demonstrated that a political economy approach is needed to move urban transport research and action on walking and cycling, beyond the observed behaviour and features, to an examination of the underlying political, social and economic structures that influence development of transport and urban planning policies that either promote or marginalize walking and cycling. A political economy helps us to dig deeper into the values, interests, power, decisions, logic and practices that shape walking and cycling policy. The mere existence of technical knowledge and solutions on walking and cycling does not guarantee implementation. The agency of the political and economic environment at local, national and international levels plays a key role in translating the knowledge into practical solutions in the form of complete streets, active design guidelines, safe walking and cycling environments, an integrated public transport system, and allocation of human and financial resources to planning for cycling and walking.

A question to conclude with is: Should transport research on walking and cycling in Africa continue to be based on the positivist approach? While a number of chapters in this book address both technical, and some of the political economy aspects, of walking and cycling in Africa, the chapter concludes that there are a number of political economy questions regarding walking and cycling in urban Africa that require further in-depth investigation, in current and future research. Beyond research, a political economy and policy analysis identifies actions that can be taken to address or influence an issue (Weible et al., 2012). It is the hope of the authors that researchers and practitioners, working on walking and cycling policy in urban Africa, will use their positions or situations to leverage different institutions, not only to provide funding for research, but also develop and implement integrated transport solutions in favour of walking, cycling and public transport. Three recent studies on institutions in urban transport (Kanyama and Cars, 2009), transport governance indicators for Sub-Saharan Africa (Christie et al., 2013) and the political economy of transportation policy and practice in Nairobi (Klopp, 2011) are encouraging pointers to take up some of the urban transport political economy questions that this chapter has highlighted. It is also encouraging to note the increasing attention being paid to urban transport research by the Institute for Development Studies of the University of Nairobi. Some of the research at this institute is looking into institutional and political factors affecting walking and cycling in particular, and urban transport policy in general (see Chapters 7 and 12 in this book).

References

Alliance for Biking and Walking. 2010. *Bicycling and walking in the U.S.: benchmarking reports.* Washington, DC: Alliance for Biking and Walking.

Ahmed, Q.I., Lu, H. and Ye, S. 2008. Urban transportation and equity: a case study of Beijing and Karachi. *Transportation Research Part A*, 42: 125–139.

Behrens, R. 2005. Accommodating walking as a travel mode in South African Cities: towards improved neighbourhood movement network design practices. *Planning Practice and Research*, 20(2): 163–182.

Black, C., Parkhurst, G. and Shergold, I. 2016. The EVIDENCE project: origins, review findings and prospects for enhanced urban transport appraisal and evaluation in the future. *Transport Policy & Practice*, 22(1 and 2): 6–225.

Boisvert, V. and Vivien, F.D. 2012. Towards a political economy approach to the convention on biological diversity. *Cambridge Journal of Economics*, 36: 1163–1179.

Christie, A., Smith, D. and Conroy, K. 2013. *Transport governance indicators for Sub-Saharan Africa.* Washington, D. C: Sub-Saharan Africa Transport Policy Program (Working Paper No. 95).

Clark, B. 1998. *Political economy: a comparative approach.* Westport: Praeger Publishers.

Department for International Development. 2009. *Political economy analysis: how to note.* A DFID practice paper.

Dimitriou, H. T. 2011. *Urban transport planning: a developmental approach.* Abington: Routledge.

Docherty, I. and Shaw, J. 2011. The transformation of transport policy in Great Britain? 'New Realism' and New Labour's decade of displacement activity. *Environment and Planning A*, 63: 224–251.

Flyvbjerg, B. 2002. Bringing power to planning: one researcher's praxis story. *Journal of Planning Education and Research*, 21(4): 353–366.

Flyvbjerg, B. 2005. Design by deception: the politics of megaproject approval. *Harvard Design Magazine*, 22: 50–59.

Flyvbjerg, B. 2007. Policy and planning for large-infrastructure projects: problems, causes, cures. *Environment and Planning B: Planning and Design*, 34: 578–597.

Flyvbjerg, B. 2009. Survival of the unfittest: why the worst infrastructure gets built – and what we can do about it. *Oxford Review of Economic Policy*, 25(3): 344–367.

Fox, A.M. and Reich, M.R. 2013. Political economy of reform. In A.S. Preker, M.E. Lindner, D. Chernichovsky and O.P. Schellekens (eds), *Scaling up affordable health insurance* (pp. 395–434). Washington, DC: World Bank.

Galeano, E. 2009. *Open veins of Latin America: five centuries of the pillage of a continent.* London: Serpent's Tail.

Garrison, W.L. and Levinson, D.M. 2006. *The transportation experience: policy, planning, and deployment.* Oxford: Oxford University Press.

Goetz, A.R., Vowles, T.M. and Tierney, S. 2009. Bridging the qualitative-quantitative divide in transport geography. *The Professional Geographer*, 61(3): 323–335.

Hook, W. 1994. *Counting on cars, counting out people: a critique of the World Bank's procedures for the transport sector and their environmental implications.* New York: Institute for Transportation and Development Policy.

Institute of Transportation Engineers. 1965. *Traffic engineering handbook.* Washington: Institute of Transportation Engineers.

Kane, L. 2011. Building the Foreshore Freeways: the politics of a freeway "artifact". South African Transport Conference. Accessed at: www.lisakane.co.za (accessed 4 April, 2014).

Kanyama, A. and Cars, G. 2009. *In search of a framework for institutional coordination in the planning for public transportation in Sub-Saharan African Cities: an analysis based on experiences from Dar-es-Salaam and Nairobi.* Stockholm: Royal Institute of Technology (Research Report).

Khayesi, M., Monheim, H., and Nebe, J. M. 2010. Negotiating "streets for all" in urban transport planning: the case for pedestrians, cyclists and street vendors in Nairobi, Kenya. *Antipode: A Radical Journal of Geography*, 42(1): 103–126.

Klopp, J.M. 2011. Towards a political economy of transportation policy and practice in Nairobi. *Urban Forum* (2012) 23: 1–21, DOI 10.1007/s12132-011-9116-y.

Knoflacher, H. 2009. From myth to science in urban and transport planning: from uncontrolled to controlled and responsible urban development in transport planning. *International Journal of Injury Control and Safety Promotion*, 16(1): 3–7.

Koglin, T. and Rye, T. 2014. The marginalisation of bicycling in modernist urban transport planning. *Journal of Transport & Health*, 1: 214–222.

Legacy, C., Curtis, C. and Sturup, S. 2012. Is there a good governance model for the delivery of contemporary transport policy and practice? An examination of Melbourne and Perth. *Transport Policy*, 19: 8–16.

Leys, C. 1975. *Under-development in Kenya: the political economy of neo-colonialism.* Nairobi: East African Educational Publishers.

Liodakis, G. 2010. Political economy, capitalism and sustainable development. *Sustainability*, 2: 2601–2616.

Litman, T. 2013. The new transportation planning paradigm. *ITE Journal*, 83(6): 20–28.

Lucas, K. and Currie, G. 2012. Developing socially inclusive transportation policy: transferring the United Kingdom Policy approach to the State of Victoria? *Transportation,* 39(1):151–173.

Mees, P. 2009. Density delusion? Urban form and sustainable transport in Australian, Canadian and US Cities. *World Transport Policy & Practice*, 15(2): 29–42.

Mees, P. 2010. Public transport policy in Australia: a density delusion? *World Transport Policy & Practice*, 16(3): 35–42.

Monheim, H. 1996. Mobility without cars. *Transportation and Environment*, 21: 31–34.

Pellegrino, G. (ed). 2011. The politics of proximity: mobility and immobility in practice. Surrey: Ashgate.

Pignataro, L. J. 1973. *Traffic engineering: theory and practice*. Englewood Cliff, New Jersey: Prentice-Hall.

Porter, G. 2014. Transport services and their impact on poverty and growth in rural Sub-Saharan Africa: a review of recent research and future research needs. *Transport Reviews*, 34(1): 25–45.

Reich, M.R. 2002. Reshaping the state from above, from within, from below: implications for public health. *Social Science & Medicine*, 54: 1669–1675.

Salon, D. and Aligula, E.M. 2012. Urban travel in Nairobi, Kenya: analysis, insights, and opportunities. *Journal of Transport Geography*, 22: 65–70.

Salon, D. and Gulyani, S. 2010. Mobility, poverty and gender: travel choices of slum residents in Nairobi, Kenya. *Transport Reviews*, 30(5): 641–657.

Stopher, P.R. and Greaves, S.P. 2007. Household travel surveys: where are we going? *Transportation Research A*, 41(5): 367–381.

Stucki, M. 2015. *Policies for sustainable accessibility and mobility in urban areas of Africa*. Washington, DC: Sub-Saharan Africa Transport Policy Program (Working Paper No. 16).

Texas Institute of Technology. 1996. *Quantifying congestion: final report*.

Transportation Research Board. 1985. *Highway Capacity Manual*. Washington, D.C.: Transportation Research Board.

Tyler, N. 2004. *Justice in transport policy*. London: Centre for Transport Studies, University College London (Working Paper 8).

Vasconcellos, E.A. 2001. *Urban transport, environment and equity: the case for developing countries*. London: Earthscan Publications.

Venter, C., Vokolkova, V. and Michalek, J. 2007. Gender, residential location, and household travel: empirical findings from low-income urban settlements in Durban, South Africa. *Transport Reviews*, 27(6): 653–677.

Wachs, M. 1995. The political context of transportation policy. In S. Hanson (ed), *The geography of urban transportation* (pp. 269–286). New York: The Guilford Press.

Weible, C.M., Heikkila, T., de Leon, P. and Sabatier, P.A. 2012. Understanding and influencing the policy process. *Policy Sciences*, 45: 1–21.

Whitelegg, J. 2011. Editorial. *World Transport Policy and Practice*, 17(3): 3–4.

Wilkinson, P. 2008. Reframing urban passenger transport provision as a strategic priority for developmental local government. In M. van Donk, M. Swilling, E. Pieterse and S. Parnell (eds), *Consolidating developmental local government: lessons from the South African experience* (pp. 203–222). Cape Town: University of Cape Town Press.

Wolfe, R. I. 1963. *Transportation and politics*. Princeton, NJ: D. Van Nostrand.

Index